Active Applied Discrete Structures

Active Applied Discrete Structures

Ken Levasseur
University of Massachusetts Lowell

May 15, 2021

Edition: version 0.1

Website: discretemath.org

©2020   Ken Levasseur

9 781008 961227

Preface

Introduction. *Active Applied Discrete Structures* is designed for use in a flipped class using *Applied Discrete Structures*. Each chapter, designed for a 75 minute class period. There are two parts that correspond with a two semester sequence, *Part 1 - Fundamentals*, and *Part 2 - Structures*.

Usage. The class format that is presumed is for the instructor to assign students to read materials that will be covered in class n in time for students to submit a short assignment by the beginning of class $n-1$. Then, at the time of class n, they work on more challenging problems in groups of 4-6 students.

A tricky thing about this format is getting started. On class 1 of the first semester, no prior reading can be expected. The binary representation of positive integers is discussed. This is mostly done through group work. The assumption is that since most students in the class are computer science majors, they have some familiarity with the topic and can help other students with the problems. Class 2 is also a bit of a rush. I ask students to do reading for that class and get responses back to me a couple of days before the second class. They are also asked to prepare for Class 3 and get responses to me by Class 2. Fortunately, the material in Chapter 1 isn't too difficult. The rest of the semester proceeds as described above. In the second semester, the first class is taken up by reviewing the most important concepts from the first semester through sequence of problems. Another quick turn-around in the reading for Class 2 is needed to get into the flow. After that the second semester proceeds the same as the first.

Contents of the Chapters. Each chapter has two or three sections. Each chapter except the first has a reading section that contains instructions to student prior the class period. Generally, this is to read one or more sections of the text, respond to a "Response Question" and do a few basic problems associated with the reading. The Response Questions tend to be more tangential to the content and often connect the mathematics to computer science.

The second section is always a series of problems to be worked on during the class. For students' convenience, some chapters have a third section, usually containing a short excerpt from *Applied Discrete Structures*.

The Problems. The problems and ideas that make up this work come from several sources. Many of the problems are taken directly from *Applied Discrete Structures*, mostly even numbered problems for which no solutions are published. In some cases, the problem are slightly altered. I've also mined problems from other sources. Some were clearly in the "public domain" in that they have appeared in several places over many years. In those cases, I haven't identified the source. There have been a few problems or ideas that seemed somewhat more novel, and I've noted their source. Books by Bogart ([1]) and Levin ([3]) are among these sources.

I've attempted to mix relatively easy problems with some more challenging ones for in-class work, while keeping the number of problems doable for most of the students.

Technology - Zoom and Piazza. Halfway into testing Part 1 of this document, the Covid-19 pandemic forced my class online. The transition wasn't perfect, but by using Zoom and Piazza I managed to approximate the face-to-face experience to some degree.

After answering questions and a short introduction to the topic of the day, I distributed students into breakout rooms in Zoom. The groups worked on the in-class problems and posted their solutions on Piazza. Within a corresponding Piazza group, they would post the answer to on of the questions. I could monitor each group's Piazza posts in real time, commenting on or endorsing the solutions. If needed, I could also drop in on the group's breakout room to help out with any difficulties. Students were brought back to the main room for the final 15 minutes or so to review the solutions that were found by the groups. Finally, I would pin correct solutions and make them available to the whole class as a record of the work done that day. This final step added value over what we would do in the face-to-face meetings.

Status. As of the spring of 2020, *Part 1 - Fundamentals*, contains 24 chapters class-tested in the first semester course at UMass Lowell. This covers the first eight chapters of *Applied Discrete Structures*. *Part 2 - Structures* will contain another 24 chapters for use in a second semester course.

The main web page for Applied Discrete Structures is http://discretemath.org

Ken Levasseur
Lowell, MA

Contents

II Structures

Back Matter

Part I

Fundamentals

Chapter 1

Binary Representation of Positive Integers

1.1 Reading Assignment

Since this is the first class meeting, there is no prior reading. Half of the class is devoted to explaining the way the class will be run. Then we will explore the binary representation of positive integers, which is in Section 1.4 of *Applied Discrete Structures*. A sheet with the base 10 numbers 1 through 64 and their corresponding binary representations is passed out. Students are asked to identify patterns.

1.2 In-class Exercises

Work on these in class

Exercises

1. Spend a few minutes looking at the tables in Section 1.3, p. 4 and make a few observations.

2. What base 10 number is equal to 101000010_2?

3. What is the base 2 representation of 911?

4. An even number is an (integer) multiple of 2. For example, 12 is even because $12 = 6 \cdot 2$ but 13 is not even since $12 = \frac{13}{2} \cdot 2$. How can you quickly tell whether a number represented in base 10 is even? How can you quickly tell whether a number represented in base 2 is even?

5. How can you quickly tell whether a number represented in base 10 is a multiple of 5? Can you quickly tell whether a number represented in base 2 is a multiple of 5?

6. How can you quickly tell whether a number represented in base 10 is a multiple of 8? Can you quickly tell whether a number represented in base 2 is a multiple of 8?

7. How can you quickly tell whether a number represented in base 10 is a multiple of 9? Can you quickly tell whether a number represented in base 2 is a multiple of 9?

1.3 Binary Conversion Tables

Look for patterns in these two tables. The second gives the binary form of integers padded with 0's so as to contain exactly 4 bits.

Base 10	Base 2	Base 10	Base 2
1	1_2	33	100001_2
2	10_2	34	100010_2
3	11_2	35	100011_2
4	100_2	36	100100_2
5	101_2	37	100101_2
6	110_2	38	100110_2
7	111_2	39	100111_2
8	1000_2	40	101000_2
9	1001_2	41	101001_2
10	1010_2	42	101010_2
11	1011_2	43	101011_2
12	1100_2	44	101100_2
13	1101_2	45	101101_2
14	1110_2	46	101110_2
15	1111_2	47	101111_2
16	10000_2	48	110000_2
17	10001_2	49	110001_2
18	10010_2	50	110010_2
19	10011_2	51	110011_2
20	10100_2	52	110100_2
21	10101_2	53	110101_2
22	10110_2	54	110110_2
23	10111_2	55	110111_2
24	11000_2	56	111000_2
25	11001_2	57	111001_2
26	11010_2	58	111010_2
27	11011_2	59	111011_2
28	11100_2	60	111100_2
29	11101_2	61	111101_2
30	11110_2	62	111110_2
31	11111_2	63	111111_2
32	100000_2	64	1000000_2

n	padded binary n
0	0000
1	0001
2	0010
3	0011
4	0100
5	0101
6	0110
7	0111
8	1000
9	1001
10	1010
11	1011
12	1100
13	1101
14	1110
15	1111

Chapter 2

Sets and Operations on them

2.1 Reading Assignment

Before class, read Sections 1.1 and 1.2 of *Applied Discrete Structures*. There is a lot of terminology and notation in these two sections that is used throughout the book. Whatever aids you personally use to remember them, by all means use them.

Reading Questions

Turn in solutions to these exercises.

1. **Response Question.** How are the set operations union and intersection similar to the operations addition and multiplication on numbers, and how are they different?

2. List all elements of the following sets:

   (a) $\{\frac{1}{n} \mid n \in \{3, 4, 5, 6\}\}$

   (b) $\{x \in \mathbb{Z} \mid x = x + 1\}$

   (c) $\{n^2 \mid n = -2, -1, 0, 1, 2\}$

   (d) $\{n \in \mathbb{P} \mid n \text{ is a factor of } 24\}$

3. Let $A = \{0, 2, 3\}$, $B = \{2, 3\}$, $C = \{1, 5, 9\}$, $D = \{3, 2\}$, and $E = \{2, 3, 2\}$. Assume that the universal set is $U = \{0, 1, 2, ..., 9\}$. Determine which of the following are true. Give reasons for your decisions.

   (a) $A = B$ (e) $A \cap B = B \cap A$

   (b) $B = C$ (f) $A \cup B = B \cup A$

   (c) $B = D$ (g) $A - B = B - A$

   (d) $E = D$ (h) $A \oplus B = B \oplus A$

2.2 In-Class Exercises

Exercises

1. Use set-builder notation to describe the following sets of positive integers:

    (a) $\{1, 2, 3, 4, 5, 6, 7\}$

    (b) $\{1, 10, 100, 1000, 10000, 100000\}$

2. Let $U = \{1, 2, 3, ..., 9\}$. Find an example to illustrate that there are sets A and B such that $A - B \neq B - A$

3. Suppose that U is an infinite universal set, and A and B are infinite subsets of U. Answer the following questions with a brief explanation.

    (a) Must A^c be finite?

    (b) Must $A \cup B$ be infinite?

    (c) Must $A \cap B$ be infinite?

4. Find two sets A and B for which $|A| = 5$, $|B| = 6$, and $|A \cup B| = 9$. What is $|A \cap B|$?

5. For any sets A and B, define $A \times B = \{(a, b) \mid a \in A \text{ and } b \in B\}$ and $AB = \{ab \mid a \in A \text{ and } b \in B\}$. If $A = \{1, 2\}$ and $B = \{2, 3, 4\}$, what is $|A \times B|$? What is $|AB|$?

6. A common data structure for a software implementation of sets is a "bitmap." The way it works is if you want to work with subsets of a universe, U, with cardinality n you first establish an ordering of U when u_k is the kth element. A set A is then represented by a string of n bits $b_1 b_2 \ldots b_n$ when b_k is 1 if $u_k \in A$ and is 0 otherwise. In the following questions, assume $U = \{1, 2, 3, 4, 5\}$ with the ordering as listed.

    (a) What are the bit strings for the empty set and for U?

    (b) What are the bit strings for $A = \{1, 2, 3\}$ and $B = \{1, 3, 5\}$?

    (c) What are the general rules for determining the the bit strings for $A \cap B$ and $A \cup B$? What their bit strings in this particular case?

Chapter 3

Sets, Sums & Products

3.1 Reading Assignment

Read Sections 1.3 and 1.5 of *Applied Discrete Structures* There are three main topics in the readings. In the next chapter, we will be counting things. Cartesian products of sets and the power set of a set are two structures that will will be used in some of the counting. Be sure to understand the distinction between the two topics. A Cartesian product involves of two or more sets and is a set of ordered pairs (or triples, quadruples,...); while a power set is of a single set and is the set of all subsets of that set. The third topic, summation notation, may very well be review for many readers. It also will appear in the counting process on several occasions..

Question 3.1.1 Response Question. If A is a finite set, why is the number of elements in the power set of A a power of 2? □

Also, turn in solutions to these exercises:

Exercises

1. Let $B = \{0, 1\}$. List elements of $\mathcal{P}(B)$, $B \times B$ and $B \times B \times B$.

2. Calculate $\sum_{k=1}^{3}(2k-1)$, $\sum_{k=1}^{4}(2k-1)$, and $\sum_{k=1}^{5}(2k-1)$. Do you see a pattern?

3.2 In-Class Exercises

1. Let $X = \{n \in \mathbb{N} \mid 10 \leq n < 20\}$. Find examples of sets with the properties below and very briefly explain why your examples work.

   (a) A set $A \subseteq \mathbb{N}$ with $|A| = 10$ such that $X - A = \{10, 12, 14\}$.

   (b) A set $B \in \mathcal{P}(X)$ with $|B| = 5$.

   (c) A set $C \subseteq \mathcal{P}(X)$ with $|C| = 5$.

   (d) A set $D \subseteq X \times X$ with $|D| = 5$.

   (e) A set $E \subseteq X$ such that $|E| \in E$.

2. (From [3]) Explain why there is no set A which satisfies $A = \{2, |A|\}$

3. Use summation or product notation to rewrite the following.

   (a) $1 + \dfrac{1}{2} + \dfrac{1}{3} + \dfrac{1}{4} + \cdots + \dfrac{1}{50}$

   (b) $1 + 5 + 9 + 13 + \cdots + 421$

   (c) $\dfrac{1}{2} \cdot \dfrac{3}{4} \cdot \dfrac{5}{6} \cdot \cdots \cdot \dfrac{99}{100}$

4. Are there sets A and B such that $|A| = |B|$, $|A \cup B| = 10$, and $|A \cap B| = 5$? Explain.

5. (from [3]) Consider the universe of positive integers greater than or equal to 2. Let A_2 be the set of all multiples of 2 except for 2. Let A_3 be the set of all multiples of 3 except for 3. And so on, so that A_n is the set of all multiple of n except for n, for any $n \geq 2$. Describe (in words) the set $(A_2 \cup A_3 \cup A_4 \cup \cdots)^c$.

Chapter 4

Counting: Product Rule and Permutations

4.1 Reading Assignment

Read Sections 2.1 and 2.2 of *Applied Discrete Structures*. You will learn about the Rule of Products, which is one of the fundamental rules of counting. In Section 2.2 we will apply to the rule of products to a very common situation, when we want to know how many ways we can put a set of objects in order.

Question 4.1.1 Response Question. Suppose A and B are finite sets. Explain how the cardinality the Cartesian product $A \times B$ can be determined using the Rule of Products. $\qquad\square$

Also, turn in solutions to these exercises:

Exercises

1. A builder of modular homes would like to impress his potential customers with the variety of styles of his houses. For each house there are blueprints for three different living rooms, four different bedroom configurations, and two different garage styles. In addition, the outside can be finished in cedar shingles or brick. How many different houses can be designed from these plans?

2. How many ways can the letters in the word DRACUT be arranged? They don't have to form a real word.

4.2 In-Class Exercises

Exercises

1. How many of the integers from 100 to 999 have the property that the sum of their digits is even? For example, 561 would be counted, but 214 would not be counted.

2. How many positive integers divide evenly into $67,500 = 2^2 3^3 5^4$?

3. The manager of a baseball team has decide on the batting order of his team. He has selected the nine batters already.

   (a) How many ways could he select a batting order?

   (b) He decides that the catcher must bat before the shortstop? How many ways can he select a batting order now?

   (c) In addition to the restriction about the catcher and shortstop, suppose he decides that the pitcher must bat immediately after the first baseman. How many ways can the manager select a batting order now?

4. How many ways can the letters in the word APPLE be arranged?

Chapter 5

Partitions and Combinations

5.1 Reading Assignment

Read Sections 2.3 and 2.4 of Applied Discrete Structures. In Section 2.3, you will learn about the Law of Addition, which is another fundamental counting principle. Finally, we turn to the problem of counting subsets of a set with a fixed size. The numbers we arrive at happen to be binomial coefficients, which you may have seen in another context, the expansion of an expression like $x + y$ to a power.

Question 5.1.1 Response Question. In mathematics, the word partition is used in two contexts. One is for partitions of sets, as described in Section 2.3. The other is for partitions of a positive integer. An example of a partition of 5 is $3 + 1 + 1$, a sum of positive integers equal to 5. It is customary to write the terms of the sum in non-increasing order since $1 + 3 + 1$ is considered the same partition of 5. The other partitions of 5 are 5, $4 + 1$, $3 + 2$, $2 + 2 + 1$, $2 + 1 + 1 + 1$, and $1 + 1 + 1 + 1 + 1$. How might a listing of all partitions of an integer like 5 help in listing all partitions of a set with that many elements?

□

Exercises to do and turn in:

Exercises

1. Which of the following collections of subsets of the plane, $\mathbb{R}^2$, are partitions?

    (a) $\{\{(x, y) \mid x + y = c\} \mid c \in \mathbb{R}\}$

    (b) The set of all circles in $\mathbb{R}^2$

    (c) The set of all circles in $\mathbb{R}^2$ centered at the origin together with the set $\{(0, 0)\}$

    (d) $\{\{(x, y)\} \mid (x, y) \in \mathbb{R}^2\}$

2. The congressional committees on mathematics and computer science are made up of five representatives each, and a congressional rule is that the two committees must be disjoint. If there are 385 members of congress, how many ways could the committees be selected?

5.2 In-Class Exercises

Exercises

1.

   (a) A group of 30 students were surveyed and it was found that 18 of them took Calculus and 12 took Physics. If all students took at least one course, how many took both Calculus and Physics? Illustrate using a Venn diagram.

   (b) What is the answer to the question in part (a) if five students did not take either of the two courses? Illustrate using a Venn diagram.

2. How many different partitions are there of the set $\{1, 2, 3, 4, 5\}$

3. How many ways can you arrange the letters in the word BOOK-KEEPER?

4. Explain in words why the following equalities are true based on number of subsets, and then verify the equalities using the formula for binomial coefficients.

   (a) $\binom{n}{1} = n$

   (b) $\binom{n}{k} = \binom{n}{n-k}$, $0 \leq k \leq n$

5. The image below shows a 6 by 6 grid and an example of a **lattice path** that could be taken from $(0, 0)$ to $(6, 6)$, which is a path taken by traveling along grid lines going only to the right and up. How many different lattice paths are there of this type? Generalize to the case of lattice paths from $(0, 0)$ to (m, n) for any nonnegative integers m and n.

6. How many of the lattice paths from $(0, 0)$ to $(6, 6)$ pass through $(2, 3)$ as the one in Figure 1, p. 12 does?

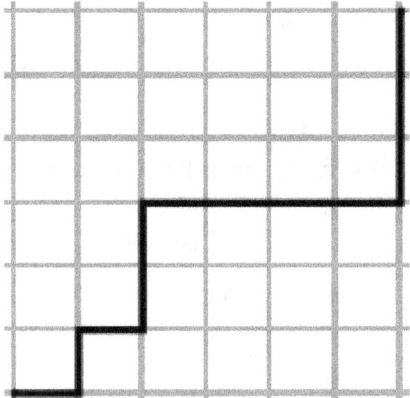

Figure 5.2.1 A lattice path

7. Consider the set of lattice paths from $(0, 0)$ to $(8, 8)$. You should know one quick formula for the cardinality of that set. However, counting a different way can lead to an interesting identity involving binomial coefficients. Notice that any path goes through exactly one of the points $(0, 8), (1, 7), (2, 6), \ldots, (8, 0)$. Count the number of lattice paths

that go through each of those 9 points - leave the expression in terms of binomial coefficients. Even more interesting is what you get if generalize to a destination of (n,n), $n \geq 1$.

5.3 Some Lattices

Here are a couple of lattices for you to doodle with.

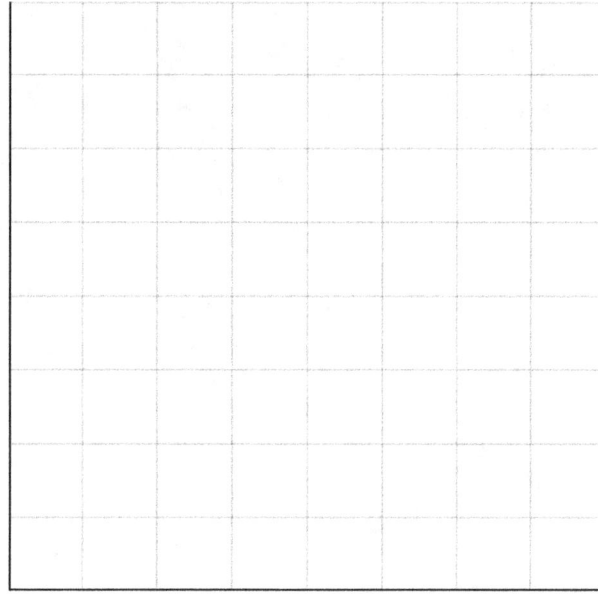

Figure 5.3.1

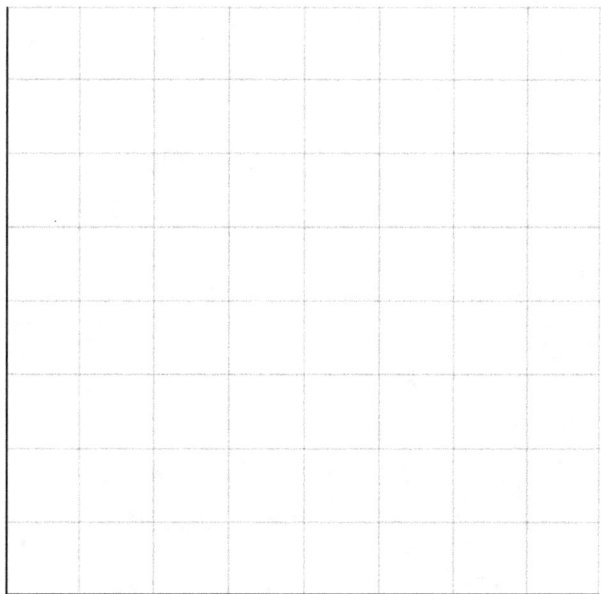

Figure 5.3.2

Chapter 6

Logic: Propositions and Truth Tables

6.1 Reading Assignment

Read sections 3.1 and 3.2 of Applied Discrete Structures. In Logic, we focus on the truth values of propositions, which are declarative sentences that are either true or false. Later we will consider propositions that involve variables that determine the truth of a proposition. But for now, we focus on how propositions can be combined to create new propositions, much as we combine numbers to get other numbers.

Question 6.1.1 Response Question. Suppose you were given a proposition generated by 100 propositional variables and you are asked whether there is at least one assignment of truth values that you could assign to these variables to make the proposition true. Why is constructing a truth table not practical. If you decided to examine all possible assignments of truth values and your computer could check one million cases per second, approximately how long would it take to check all cases? □

Also, turn in solutions to these exercises:

Exercises

1. For each of the following propositions, identify simple propositions, express the compound proposition in symbolic form, and determine whether it is true or false:

   (a) The world is flat or zero is an even integer.

   (b) If 432,802 is a multiple of 4, then 432,802 is even.

   (c) 5 is a prime number and 6 is not divisible by 4.

   (d) $3 \in \mathbb{Z}$ and $3 \in \mathbb{Q}$.

   (e) $2/3 \in \mathbb{Z}$ and $2/3 \in \mathbb{Q}$.

   (f) The sum of two even integers is even and the sum of two odd integers is odd.

2. Construct the truth tables of:

(a) $\neg(p \wedge q)$ (b) $(p \wedge q) \wedge r$

6.2 In-Class Exercises

Exercises

1. Reword the following statements into "If...then" statements.

(a) No resident of Chelmsford likes hot peppers.

(b) For 3+7=10, it is necessary that cows fly.

(c) For 3+7=10, it is sufficient that cows fly.

(d) Lowell is the oldest city in Massachusetts unless mermaids exist.

(e) I carry an umbrella when it rains.

2. Construct the truth table for $(p \vee q) \wedge (p \vee \neg q)$. Notice anything about the result?

3. Consider the statement "If Boris visits Hampton Beach, then he eats fried clams."

(a) Write the converse of the statement.

(b) Write the contrapositive of the statement.

(c) Is it possible for the contrapositive to be false? If it was, what would that tell you?

(d) Suppose the original statement is true, and that Boris eats fried clams. Can you conclude anything (about his travels)?

(e) Suppose the original statement is true, and that Boris does not eat fried clams. Can you conclude anything (about his travels)?

4. Consider the statement, "If a number is triangular or square, then it is not prime"

(a) Make a truth table for the statement $(T \vee S) \to \neg P$.

(b) If you believed the statement was false, what properties would a counterexample need to possess? Explain by referencing your truth table.

(c) If the statement were true, what could you conclude about the number 5657, which is definitely prime? Again, explain using the truth table.

Chapter 7

Equivalence, Implication, and Laws of Logic

7.1 Reading Assignment

Read sections 3.3 and 3.4 of Applied Discrete Structures. Two propositions may be totally unrelated. For example, "It snowed today." and "I played chess today." are likely to be unrelated. Yet "It snowed today." and "I played golf today." are two propositions that are related in that it is unlikely that both are true. In these two sections we introduce the ideas of equivalence and implication, which give us a precise way to talk about how propositions are related.

Question 7.1.1 Response Question. Explain why every proposition implies a tautology. □

Also, turn in solutions to these exercises:

Exercises

1.
   (a) Construct the truth table for $x = (p \wedge \neg q) \vee (r \wedge p)$.

   (b) Find an example other than x itself of a proposition generated by p, q, and r that is equivalent to x.

   (c) Find an example of a proposition that is not equivalent to x or a contradiction that implies x.

   (d) Find an example of a proposition that is not equivalent to x or a tautology that is implied by x.

2. Show that the common fallacy $(p \rightarrow q) \wedge \neg p \Rightarrow \neg q$ is not a law of logic.

7.2 In-Class Exercises

Exercises

1. Find a proposition that is equivalent to $p \vee q$ and uses only conjunction and negation.

2. Frankie Fib was telling you what he consumed yesterday afternoon. He tells you, "I had either popcorn or raisins. Also, if I had cucumber sandwiches, then I had soda. But I didn't drink soda or tea." Of course you know that Frankie is the worlds worst liar, and everything he says is false. What did Frankie have to eat and drink?

3. Construct the truth table for $(p \to q) \wedge (q \to r) \wedge (r \to p)$. Notice anything about the result?

4. The significance of the Sheffer Stroke is that it is a "universal" operation in that all other logical operations can be built from it.

   (a) Prove that $p|q$ is equivalent to $\neg(p \wedge q)$.

   (b) Prove that $\neg p \Leftrightarrow p|p$.

   (c) Build $\wedge$ using only the Sheffer Stroke.

   (d) Build $\vee$ using only the Sheffer Stroke.

7.3 The Sheffer Stroke

Another logical operation is the Sheffer Stroke, which is the subject of one of the exercises.

Table 7.3.1 Truth Table for the Sheffer Stroke

p	q	$p \mid q$
0	0	1
0	1	1
1	0	1
1	1	0

Chapter 8

Structured Proofs

8.1 Reading Assignment

Read section 3.5 of Applied Discrete Structures. We can say that there are various relationships between propositions, but it might not be obvious that what we say is really true. A proof is meant to be a means by which one can be convinced that such a relationship is true. Proofs are central to mathematics and appear in all remaining chapters in this book. In Section 3.5 we start by identifying two basic types of proof: direct and indirect.

Question 8.1.1 Response Question. A proposition, P, generated by a set of propositional variables is said to be satisfiable if there is at least one way to assign truth values to all of the variables so that P Is true. Explain why P is satisfiable as long as $\neg P$ is not a tautology. ☐

Also, turn in solutions to these exercises.

Exercises

1. Put the following into symbolic form and check its validity: If I am a good person, nothing bad will happen to me. Nothing happened to me. Therefore, I am a good person.

2. Give a direct or indirect proof of:

$$p \to q, \neg r \to \neg q, \neg r \Rightarrow \neg p$$

8.2 In-Class Exercises

Exercises

1. Prove either directly or indirectly:

$$a \lor b, c \land d, a \to \neg c \Rightarrow b$$

2. In these two Lewis Carroll puzzles, you are given premises and are expected to form your own conclusion. In each of them, convert the premises to symbolic form, draw a conclusion, and then translate back

to English.

(a) • No bald creature needs a hairbrush.

   • No lizards have hair.

(b) • Promise breakers are untrustworthy.

   • Wine drinkers are very communicative.

   • A man who keeps his promises is honest.

   • No teetotalers are pawnbrokers.

   • One can always trust a very communicative person.

3.   There are $n + 1$, $n \geq 1$ people who want to go to a concert. All have different ages. You have three tickets: a back-stage pass and two regular (but distinguishable) tickets. Here are the rules for passing out the tickets:

   • The backstage pass must go to the oldest person who gets a ticket.

   • The person who gets the backstage pass can' t get either of the other two tickets, but the two regular tickets can both go to the same person.

How many ways can you give away the tickets? There are two ways to count. Find both and equate them.

8.3 Basic Logical Inferences

From section 3.4 of Applied Discrete Structures:

Table 8.3.1 Basic Logical Laws - Common Implications and Equivalences

Detachment (AKA Modus Ponens)	$(p \rightarrow q) \wedge p \Rightarrow q$
Indirect Reasoning (AKA Modus Tollens)	$(p \rightarrow q) \wedge \neg q \Rightarrow \neg p$
Disjunctive Addition	$p \Rightarrow (p \vee q)$
Conjunctive Simplification	$(p \wedge q) \Rightarrow p$ and $(p \wedge q) \Rightarrow q$
Disjunctive Simplification	$(p \vee q) \wedge \neg p \Rightarrow q$ and $(p \vee q) \wedge \neg q \Rightarrow p$
Chain Rule	$(p \rightarrow q) \wedge (q \rightarrow r) \Rightarrow (p \rightarrow r)$
Conditional Equivalence	$p \rightarrow q \Leftrightarrow \neg p \vee q$
Biconditional Equivalences	$(p \leftrightarrow q) \Leftrightarrow (p \rightarrow q) \wedge (q \rightarrow p) \Leftrightarrow (p \wedge q) \vee (\neg p \wedge \neg q)$
Contrapositive	$(p \rightarrow q) \Leftrightarrow (\neg q \rightarrow \neg p)$

Chapter 9

Mathematical Induction

9.1 Reading Assignment

Read Sections 3.6 and 3.7 of *Applied Discrete Structures*. It is only necessary to read 3.6 through Example 3.6.7. The main idea in this reading is Mathematical Induction, which is a proof technique for propositions that have an integer variable. This type of proof is used throughout mathematics and is particularly prevalent in discrete settings.

Question 9.1.1 Response Question. You don't need induction to prove that the sum of the first n Positive integers equals $\frac{n(n+1)}{2}$. Google "Gauss sum of consecutive integers" and read about how you can do it even more simply. Explain what you read. □

Also, turn in solutions to these exercises.

Exercises

1. Simplify the expressions

    (a) $\left(\sum_{k=1}^{n+1} k^2\right) - \left(\sum_{k=1}^{n} k^2\right)$

    (b) $\sum_{k=1}^{n}\left(\frac{1}{k} - \frac{1}{k+1}\right)$

    (c) $\frac{(n+2)!}{n!}$

2. Prove that for $n \geq 0$, $\sum_{k=0}^{n} 2^k = 2^{n+1} - 1$.

9.2 In-Class Exercises

Exercises

1.   Prove that for $n \geq 1$,

$$\frac{1}{1 \cdot 2} + \frac{1}{2 \cdot 3} + \cdots + \frac{1}{n(n+1)} = \frac{n}{n+1}.$$

2.   Prove that it is possible to make up any postage of 28 cents or more using only five-cent and eight-cent stamps.

3.   Suppose that a particular real number x has the property that $x + \frac{1}{x}$ is an integer. Prove that $x^n + \frac{1}{x^n}$ is an integer for all natural numbers n.

Chapter 10

Quantifiers and Proof Review

10.1 Reading Assignment

Read Sections 3.8 and 3.9 of *Applied Discrete Structures*. Quantifiers are used to say something about the truth sets of propositions. You will read about two basic ones (Universal and Existential). Individually, they are fairly simple. Where things can be tricky is when when you have more than one variable and you mix the two types of quantifiers, or you negate a quantified expression. Pay close attention to these situations.

Question 10.1.1 Response Question. In reviewing a certain local coffee roaster, a writer stated "...but all of its coffee is not fair trade." The writer was rebutting a claim by the roaster that "All of our coffee is fair trade." Explain why the reviewer's statement was incorrect. □

Also, turn in solutions to these exercises:

Exercises

1. Let $M(x)$ be "x is a mammal," let $A(x)$ be "x is an animal," and let $W(x)$ be "x is warm-blooded."

   (a) Translate into a formula: Every mammal is warm-blooded.

   (b) Translate into English: $(\exists x)(A(x) \wedge (\neg M(x)))$.

2. Write out a complete proof that if n is an integer, n^2 is even if and only if n is even.

10.2 In-Class Exercises

Exercises

1. Translate the following statement over the positive integers into symbols. Use $E(x)$ for "x is even" and $O(x)$ for "x is odd" in the first three parts.

   (a) No number is both even and odd.

    (b) One more than any even number is an odd number.

    (c) There is prime number that is even.

    (d) Between any two numbers there is a third number.

    (e) There is no number between a number and one more than that number.

2. Use quantifiers to state that for every positive integer, there is a larger positive integer.

3. One of the following is true and the other is false. Identify the true one says and explain why the other one is false.

$$(\exists b)_{\mathbb{Z}}((\forall a)_{\mathbb{Z}}(a + b = 0))$$
$$(\forall a)_{\mathbb{Z}}((\exists b)_{\mathbb{Z}}(a + b = 0))$$

4. Prove that the sum of of an odd integer and and even integer is odd.

5. Prove that if you divide 4 into a perfect square, $1, 4, 9, 16, \ldots$, the remainder will be either 0 or 1.

6. Prove that the cube root of 2 is an irrational number.

Chapter 11

Set Theory Logic

11.1 Reading Assignment

Read Sections 4.1 and 4.2 of *Applied Discrete Structures*. In these sections, we revisit some of the set theory from Chapter 1, applying some of the logic in Chapter 3. Be aware of the types of objects that are in the sets involved in a proof. One example is that when you are proving something about Cartesian products, the objects usually need to be identified as ordered pairs, not simply as single variables.

Question 11.1.1 Response Question. Compare the Laws of Set Theory in Section 4.2 of Applied Discrete Structures with the Basic Laws of Logic in Section 3.5 of Applied Discrete Structures. Focus on any two different laws of set theory that you choose and discuss how they are similar to two logic laws. □

Also, turn in solutions to these exercises:

Exercises

1. Write the converse of the following true statements and prove or disprove them.

   (a) Let A, B, and C be sets. If $A \subseteq B$ and $B \subseteq C$, then $A \subseteq C$.

   (b) Let A, B, and C be sets. If $(A \subseteq B$ and $A \subseteq C)$ then $A \subseteq B \cap C$.

   (c) Let A, B, and C be sets with $C \neq \emptyset$. If $A \subseteq B$ then $A \times C \subseteq B \times C$.

2.

   (a) Prove the Identity Law for sets with a membership table.

   (b) Prove the Involution Law for sets using basic definitions.

11.2 In-Class Exercises

Exercises

1. What can one say about the sets A and B if we know the following? Back up your answers with proofs.

   (a) $A \cup B = A$

   (b) $A \cap B = A$

   (c) $A - B = A$

   (d) $A \cap B = B \cap A$

   (e) $A - B = B - A$

2.

   (a) Given the following sets of integers, A, B, C, find the set of elements that belong to exactly one of the three sets.

   $$A = \{2, 6, 10, 14, 18\}$$
   $$B = \{2, 3, 5, 7, 11, 13, 17, 19\}$$
   $$C = \{3, 6, 9, 12, 15, 18\}$$

   (b) Prove that for any three sets, A, B, C,

   $$(A \cup B \cup C) \cap ((A^c \cap B^c) \cup (A^c \cap C^c) \cup (B^c \cap C^c))$$

   is the set of all elements that belong to exactly one of the three sets. Verify this fact first with the example in the previous part, where you assume that the universe is $\{1, 2, 3, \ldots, 18, 19\}$.

   (c) Find a similar expression for the set of elements that belong to exactly one of any four sets A, B, C, D.

3. Recall that the power set of any set A is the set of all subsets of A and is denoted $\mathcal{P}(A)$. Which of the following are true?

   $$\mathcal{P}(A \cap B) = \mathcal{P}(A) \cap \mathcal{P}(B)$$
   $$\mathcal{P}(A \cup B) = \mathcal{P}(A) \cup \mathcal{P}(B)$$

   If either is not true, can you replace the equals sign with $\subseteq$ or $\supseteq$ to get a true statement?

11.3 The Basic Laws of Set Theory

Table 11.3.1 Basic Laws of Set Theory

Commutative Laws	
(1) $A \cup B = B \cup A$	(1') $A \cap B = B \cap A$
Associative Laws	
(2) $A \cup (B \cup C) = (A \cup B) \cup C$	(2') $A \cap (B \cap C) = (A \cap B) \cap C$
Distributive Laws	
(3) $A \cap (B \cup C) = (A \cap B) \cup (A \cap C)$	(3') $A \cup (B \cap C) = (A \cup B) \cap (A \cup C)$
Identity Laws	
(4) $A \cup \emptyset = \emptyset \cup A = A$	(4') $A \cap U = U \cap A = A$
Complement Laws	
(5) $A \cup A^c = U$	(5') $A \cap A^c = \emptyset$
Idempotent Laws	
(6) $A \cup A = A$	(6') $A \cap A = A$
Null Laws	
(7) $A \cup U = U$	(7') $A \cap \emptyset = \emptyset$
Absorption Laws	
(8) $A \cup (A \cap B) = A$	(8') $A \cap (A \cup B) = A$
DeMorgan's Laws	
(9) $(A \cup B)^c = A^c \cap B^c$	(9') $(A \cap B)^c = A^c \cup B^c$
Involution Law	
(10) $(A^c)^c = A$	

Chapter 12

Minsets and Duality

12.1 Reading Assignment

Read Sections 4.3 and 4.4 of *Applied Discrete Structures*. Minsets help answer the question of what sets can be created if you start with a certain collection of sets and apply basic set operations to them. Duality is an organizing principle that set theory shares with logic and a few other mathematical structures.

Question 12.1.1 Response Question. To what extend is there any duality in arithmetic of numbers with addition and multiplication? How does it break down where it doesn't in set theory? □

Also, turn in solutions to these exercises:

Exercises

1. Consider the subsets $A = \{1, 3, 5\}$, $B = \{2, 3, 4\}$, where $U = \{1, 2, 3, 4, 5\}$. List the nonempty minsets generated by A and B.

2. What is the dual of $A \cap (B \cap (A \cap B)^c) = \emptyset$?

12.2 In-Class Exercises

Exercises

1. A common way to denote a particular minset generated by a collection of subsets is as follows. If there are k subsets, $B_1, B_2, \ldots, B_k$, and $b = b_1 b_2 \cdots b_k$ is any string of k bits, then

$$M_b = M_{b_1 b_2 \cdots b_k} = D_1 \cap D_2 \cap \cdots \cap D_k,$$

where D_i is either B_i or B_i^c. If $b_i = 1$ then $D_i = B_i$ and if $b_i = 0$ then $D_i = B_i^c$. For example, if $k = 4$, $M_{0110} = B_1^c \cap B_2 \cap B_3 \cap B_4^c$.

(a) Suppose $U = \{1, 2, 3, 4, 5\}$, $k = 2$, $B_1 = \{1, 2\}$, and $B_2 = \{2, 3, 4\}$. List the nonempty minsets generated by B_1 and B_2 using "M_b" notation. Notice that they form a partition of U.

(b) How does this notation make help us see how many distinct nonempty minsets there could be that are generated by k subsets of a universe.

2.

(a) Partition $\{1, 2, \ldots, 8\}$ into the nonempty minsets generated by $B_1 = \{1, 2\}$, $B_2 = \{1, 3, 5, 8\}$, and $B_3 = \{2, 3, 4, 6\}$.

(b) How many different subsets of $\{1, 2, \ldots, 8\}$ can you create using B_1, B_2, and B_3 with the standard set operations?

(c) Do there exist subsets C_1, C_2, C_3 with which you can generate every subset of $\{1, 2, ..., 8\}$? If so, can you find such a collection of subsets? If not, why? You might find the Venn diagram below useful for thinking about this problem.

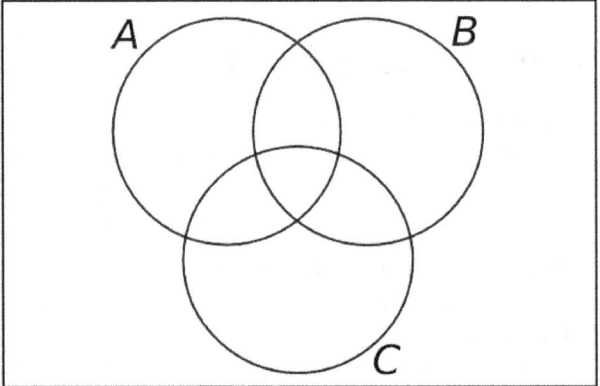

Figure 12.2.1 A three set Venn diagram

3. What is the dual of a minset? These sets are called "maxsets" Find the maxsets generated by the two sets in part (a) of the first problem. Why do you suppose they are called maxsets?

4. The descriptions of duality in Section 4.4 is not complete. If you expand expressions involving subsets, such as the expression $A \cap B \subseteq A$, which is a true statement in set theory. What should be the dual? How should we treat the subset symbol?

Chapter 13

Matrix Operations

13.1 Reading Assignment

Read Sections 5.1 and 5.2 of Applied Discrete Structures. Matrix algebra serves as a contrasting system to logic and set theory but is also needed in the study of relations in Chapter 6. In these two sections, we review matrix operations and some special matrices.

Question 13.1.1 Response Question. Let $A = \begin{pmatrix} 2 & 0 \\ 0 & -1 \end{pmatrix}$. Select any 2 by 2 matrix with nonzero entries and call it B. Compute the products AB and BA What effect does A have on B in each case? □

Also, turn in solutions to these exercises:

Exercises

1. Let $A = \begin{pmatrix} 1 & -1 \\ 2 & 3 \end{pmatrix}$ and $B = \begin{pmatrix} 0 & 1 \\ 3 & -5 \end{pmatrix}$

    (a) Compute AB and BA.

    (b) Compute $A + B$ and $B + A$.

2. For the given matrices A find A^{-1} if it exists and verify that $AA^{-1} = A^{-1}A = I$. If A^{-1} does not exist explain why.

    (a) $A = \begin{pmatrix} 2 & -1 \\ -1 & 2 \end{pmatrix}$

    (b) $A = \begin{pmatrix} 2 & 1 \\ 4 & 2 \end{pmatrix}$

There is a short video on matrix multiplication at `https://youtu.be/zt-IU1lXFzs`

13.2 In-Class Exercises

Exercises

1. Let $A = \begin{pmatrix} 1 & a \\ 0 & 1 \end{pmatrix}$ and $B = \begin{pmatrix} 1 & b \\ 0 & 1 \end{pmatrix}$. Compute the product AB.
    Based on this result, what is A^{-1}.

2. If A is a an $m \times n$ matrix, we define the transpose of A to be the $n \times m$ matrix whose rows are the columns of A. For example, the transpose of

$$\begin{pmatrix} 1 & 2 & 3 \\ 4 & 5 & 6 \end{pmatrix} \text{ is } \begin{pmatrix} 1 & 4 \\ 2 & 5 \\ 3 & 6 \end{pmatrix}.$$

The notation A^t is used for the transpose of A.

   (a) If A is an $m \times n$ matrix, are the products AA^t and $A^t A$ defined? What are the orders of the products that are defined?

   (b) Given the following matrix, what useful information might you get from the products AA^t or $A^t A$.?

$$A = \begin{pmatrix} 16 & 11 & 4 & 3 & 15 \\ 16 & 17 & 13 & 12 & 6 \end{pmatrix}$$

3. Prove by induction that for $n \geq 1$, $\begin{pmatrix} a & 0 \\ 0 & b \end{pmatrix}^n = \begin{pmatrix} a^n & 0 \\ 0 & b^n \end{pmatrix}$.

4. In this exercise, we propose to show how matrix multiplication is a natural operation. Suppose a bakery produces bread, cakes and pies every weekday, Monday through Friday. Based on past sales history, the bakery produces various numbers of each product each day, summarized in the 5×3 matrix D. It should be noted that the order could be described as "number of days by number of products." For example, on Wednesday (the third day) the number of cakes (second product in our list) that are produced is $d_{3,2} = 4$.

$$D = \begin{pmatrix} 25 & 5 & 5 \\ 14 & 5 & 8 \\ 20 & 4 & 15 \\ 18 & 5 & 7 \\ 35 & 10 & 9 \end{pmatrix}$$

   The main ingredients of these products are flour, sugar and eggs. We assume that other ingredients are always in ample supply, but we need to be sure to have the three main ones available. For each of the three products, The amount of each ingredient that is needed is summarized in the 3×3, or "number of products by number of ingredients" matrix P. For example, to bake a cake (second product) we need $P_{2,1} = 1.5$ cups of flour (first ingredient). Regarding units: flour and sugar are given in cups per unit of each product, while eggs are given in individual eggs per unit of each product.

$$P = \begin{pmatrix} 2 & 0.5 & 0 \\ 1.5 & 1 & 2 \\ 1 & 1 & 1 \end{pmatrix}$$

   These amounts are "made up", so don't used them to do your own baking!

   (a) How many cups of flour will the bakery need every Monday?

Pay close attention to how you compute your answer and the units of each number.

(b) How many eggs will the bakery need every Wednesday?

(c) Compute the matrix product DP. What do you notice?

(d) Suppose the costs of ingredients are $0.12 for a cup of flour, $0.15 for a cup of sugar and $0.19 for one egg. How can this information be put into a matrix that can meaningfully be multiplied by one of the other matrices in this problem?

Chapter 14

Matrix Laws and Oddities

14.1 Reading Assignment

Read Sections 5.3 and 5.4 of *Applied Discrete Structures*. In these two sections we concentrate on the algebraic laws of set theory and some of the laws that are missing (because they are false!).

Question 14.1.1 Response Question. Compare Matrix Law (15), The Inverse of Product Rule, with the fact that although you put your socks on before your shoes, you take your shoes off before taking off your socks.   □

Also, turn in solutions to these exercises:

Exercises

1.   Let $A = \begin{pmatrix} 0 & 1 & 0 & 0 \\ 0 & 0 & 1 & 0 \\ 0 & 0 & 0 & 1 \\ 1 & 0 & 0 & 0 \end{pmatrix}$. Compute A^2, A^3, A^4, and A^{-1}.

2.   Find at least three 2×2 matrices, A, such that $A^2 = A$.

14.2 In-Class Exercises

Exercises

1.   Let A and B be $n \times n$ matrices of real numbers. Is $A^2 - B^2 = (A - B)(A + B)$? Explain.

2.   Write each of the following systems in the form $AX = B$, and then solve the systems using matrices.
     (a) $\begin{aligned} 4x_1 - 6x_2 &= 20 \\ 3x_1 + 5x_2 &= -6 \end{aligned}$
     (b) $\begin{aligned} 5x_1 - 1x_2 &= 11 \\ -16x_1 + 5x_2 &= 12 \end{aligned}$

3.   Suppose that A, P, and B are all $m \times m$ matrices, $m \geq 2$, and $A = P^{-1}BP$. Prove that $A^n = P^{-1}B^n P$ for all $n \geq 1$.

4. Let $M_{n \times n}(\mathbb{R})$ be the set of real $n \times n$ matrices. Let $P \subseteq M_{n \times n}(\mathbb{R})$ be the subset of matrices defined by $A \in P$ if and only if $A^2 = A$. Let $Q \subseteq P$ be defined by $A \in Q$ if and only if $\det A \neq 0$.

(a) Determine the cardinality of Q.

(b) Consider the special case $n = 2$ and prove that a sufficient condition for $A \in P \subseteq M_{2 \times 2}(\mathbb{R})$ is that A has a zero determinant (i.e., A is singular) and $tr(A) = 1$ where $tr(A) = a_{11} + a_{22}$ is the sum of the main diagonal elements of A.

(c) Is the condition of part b a necessary condition?

Chapter 15

Relations

15.1 Reading Assignment

Read Sections 6.1 and 6.2 of *Applied Discrete Structures*. Relations are propositions between two variables that are boolean - they are either true or false. You've used them many times in the past without necessarily calling them by that name. For example, "less than" is a relation on the integers. $3 < 6$ is true, while $6 < 3$ is false. If we imagine all pairs of integers, (a, b), that make $a < b$ true, that set is identified as the relation "less than".

Question 15.1.1 Response Question. Although any subset of a cartesian product of a set with itself can be a relation on that set, in the long run we are most concerned with a few important ones. Three examples of very important relations are

- Less than or equal to,$\leq$, on the integers,

- Set containment, $\subseteq$, on the power set of a set,

- Logical implication, $\Rightarrow$, on any set of propositions.

Discuss any similarities you see between these three relations. $\square$

Also, turn in solutions to these exercises:

Exercises

1. Consider the two relations on people: M, where aMb if a's mother is b; and S, where aSb if a and b are siblings. Describe, in words, the two relations MS and SM.

2. Let $A = \{1, 2, 3, 4, 6, 12\}$. Draw a digraph for the relation "divides" on A.

15.2 In-Class Exercises

Exercises

1. Let S be the set of "spaces" in the floor of your classroom. Draw a digraph of the relation c, where $s_1 c s_2$ if and only if s_1 is connected to s_2 with at least one doorway.

2. Given s and t, relations on $\mathbb{Z}$, $s = \{(1, n) : n \in \mathbb{Z}\}$ and $t = \{(n, 1) : n \in \mathbb{Z}\}$, what are st and ts? Hint: Even when a relation involves infinite sets, you can often get insights into them by drawing partial graphs.

3. Let A be the set of strings of 0's and 1's of length 3 or less. This includes the empty string, λ, which is the only string of length zero.

    (a) Define the relation of w on A by xwy if x has the same number of 1's as y. For example, $01w100$, but $01w101$ is false. Draw a digraph for this relation.

    (b) Do the same for the relation p defined by xpy if x is a prefix of y. For example, $10p101$, but $01p101$ is false.

4. Consider logical implication, $\Rightarrow$, on the set of propositions $\{0, 1, p, q, p \vee q, p \wedge q, p \wedge p\}$. Draw a digraph of this relation.

Chapter 16

Properties of Relations

16.1 Reading Assignment

Read Section 6.3 of *Applied Discrete Structures*. You will read about four properties that are satisfied by certain relations. Two combinations of these properties, when true, characterize relations that are particularly important. They are partial orderings and equivalence relations. Make an effort to memorize the terms in this section - they will appear throughout the rest of the book.

Question 16.1.1 Response Question. Recall that in geometry, two triangles are similar if and only if their corresponding angles have the same measure. What kind of relation is this on the set of all triangles on the plane? □

Also, turn in solutions to these exercises:

Exercises

1. Prove that congruence modulo m is a transitive relation on the set of integers. Do this by assuming that $a \equiv_m b$ and $b \equiv_m c$, and applying the definition for $\equiv_m$ to conclude that $a \equiv_m c$.

2. Draw the ordering diagram for the relation "divides" on the divisors of $40 = 2^3 \cdot 5$.

16.2 In-Class Exercises

Exercises

1. Let $A = \{a, b, c, d\}$. Draw the graphs of relations on A where:

    (a) The first relation is reflexive, symmetric, but not transitive.

    (b) The second relation is transitive, but not symmetric and not reflexive.

    (c) The third relation is both an equivalence relation and a partial ordering.

2.   Let $A = \{0, 1, 2, 3\}$ and let

$$r = \{(0,0), (1,1), (2,2), (3,3), (1,2), (2,1), (3,0), (0,3)\}$$

(a) Verify that r is an equivalence relation on A.

(b) Let $a \in A$ and define $c(a) = \{b \in A \mid arb\}$. $c(a)$ is called the **equivalence class of a under** r. Find $c(a)$ for each element $a \in A$.

(c) Show that $\{c(a) \mid a \in A\}$ forms a partition of A for this set A.

(d) Let r be an equivalence relation on an arbitrary set A. Prove that the set of all equivalence classes under r constitutes a partition of A.

3.   Describe the equivalence classes under the relation congruence modulo 10 on the integers.

4.   Let A be the set of strings of 0's and 1's of length 3 or less; and let B be the set of strings of 0's and 1's of length 3. What properties do the following relations have?

(a) Define the relation of w on A by xwy if x has the same number of 1's as y. For example, $01w100$, but $01w101$ is false.

(b) Define the the relation d on B defined by xdy if x differs from y in exactly one position. For example, $100d101$, but $100d111$ is false.

(c) Define the the relation c defined on A by xcy if x is contained within y. For example, $10c101$, but $11c101$ is false.

For any of these relations that are partial orderings, draw the Hasse diagram for that relation. For any of them that is an equivalence relation, identify the equivalence classes.

16.3 Congruence Modulo n

This is a fundamental relation on the set of integers.

Definition 16.3.1 Congruence Modulo m. Let m be a positive integer, $m \geq 2$. We define **congruence modulo m** to be the relation $\equiv_m$ defined on the integers by

$$a \equiv_m b \Leftrightarrow m \mid (a - b)$$

$\Diamond$

Chapter 17

Relation Matrices and Closure

17.1 Reading Assignment

Read Sections 6.4 and 6.5 of *Applied Discrete Structures*. Although some relations can be computed (e. g. determining whether $a < b$), some are more efficiently represented in a matrix. We introduce the matrix representation of relations in Section 6.4, and then transitive closure in Section 6.5.

Question 17.1.1 Response Question. Let p be the relation on people where xpy if y is either x's mother or father. What is $\{z \mid xp^+z\}$, where p^+ is the transitive closure of p. □

Also, turn in solutions to this exercise:

Exercises

1. Consider the relation, s, defined by the graph in Figure 17.1.2, p. 42.

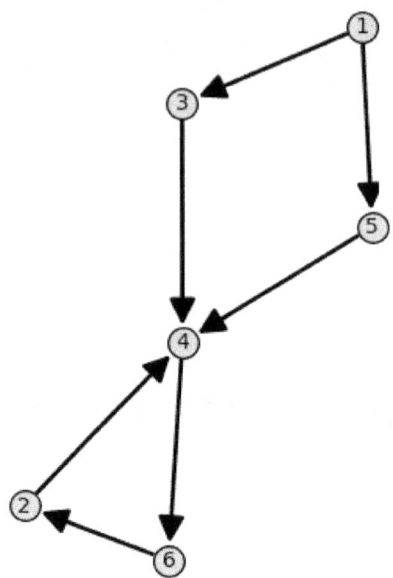

Figure 17.1.2 Digraph of s

(a) Determine the adjacency matrix of s.

(b) Use the matrix you have constructed to find the matrix of s^2.

(c) Draw the graph defined by the matrix product and verify that it is the graph of s^2.

(d) Determine the matrix of the transitive closure of s.

17.2 In-Class Exercises

Exercises

1. Let D be the set of weekdays, Monday through Friday, let W be a set of employees $\{1, 2, 3\}$ of a tutoring center, and let V be a set of computer languages for which tutoring is offered, $\{A(PL), B(asic), C(++), J(ava), L(isp), P(yt$ We define s (schedule) from D into W by dsw if w is scheduled to work on day d. We also define r from W into V by wrl if w can tutor students in language l. If s and r are defined by matrices

$$S = \begin{array}{c} \\ M \\ T \\ W \\ R \\ F \end{array} \begin{array}{ccc} 1 & 2 & 3 \\ \left(\begin{array}{ccc} 1 & 0 & 1 \\ 0 & 1 & 1 \\ 1 & 0 & 1 \\ 0 & 1 & 0 \\ 1 & 1 & 0 \end{array}\right) \end{array} \quad \text{and } R = \begin{array}{c} \\ 1 \\ 2 \\ 3 \end{array} \begin{array}{cccccc} A & B & C & J & L & P \\ \left(\begin{array}{cccccc} 0 & 1 & 1 & 0 & 0 & 1 \\ 1 & 1 & 0 & 1 & 0 & 1 \\ 0 & 1 & 0 & 0 & 1 & 1 \end{array}\right) \end{array}$$

(a) compute SR using Boolean arithmetic and give an interpretation of the relation it defines, and

    (b) compute SR using regular arithmetic and give an interpretation of what the result describes.

2.   Let $A = \{a, b, c, d\}$. Let r be the relation on A with adjacency matrix

$$
\begin{array}{c}
 \\
a \\
b \\
c \\
d
\end{array}
\begin{array}{cccc}
a & b & c & d \\
\end{array}
\left(
\begin{array}{cccc}
1 & 0 & 0 & 0 \\
0 & 1 & 0 & 0 \\
1 & 1 & 1 & 0 \\
0 & 1 & 0 & 1
\end{array}
\right)
$$

    (a) Explain why r is a partial ordering on A.

    (b) Draw its Hasse diagram.

3.   What common relations on $\mathbb{Z}$ are the transitive closures of the following relations?

    (a) aSb if and only if $a + 1 = b$.

    (b) aRb if and only if $|a - b| = 2$.

4.

    (a) Prove that if r is a transitive relation on a set A, then $r^2 \subseteq r$.

    (b) Find an example of a transitive relation for which $r^2 \neq r$.

Chapter 18

Functions and Their Properties

18.1 Reading Assignment

Read Sections 7.1 and 7.2 of *Applied Discrete Structures*. You may think you know about functions if you've taken calculus, but the approach to functions in Chapter 7, which is the more formal approach, is different. So make an effort to memorize the terms! The good news is that there are only two key properties that are introduced in Section 7.2.

Question 18.1.1 Response Question. In programming, a *function* is a named section of a program that performs a specific task and returns a value. How does this compare with the definition of a function in mathematics? □

Also, turn in solutions to these exercises:

Exercises

1. At the end of the semester a teacher assigns letter grades to each of her 45 students. Is this a function? If so, what sets make up the domain and codomain, and is the function injective, surjective, bijective, or neither?

2. Let A be a set and let S be any subset of A. Let $\chi_S : A \to \{0, 1\}$ be defined by

$$\chi_S(x) = \begin{cases} 1 & \text{if } x \in S \\ 0 & \text{if } x \notin S \end{cases}$$

   The function χ_S is called the **characteristic function** of S. Suppose $A = \{a, b, c\}$.

   (a) If $S = \{a, b\}$, list the elements of χ_S .

   (b) What are $\chi_\emptyset$ and χ_A?

18.2 In-Class Exercises

Exercises

1. Define functions on the positive integers, $\mathbb{P}$, if they exist, that have the properties specified below.

    (a) A function that is one-to-one and onto.

    (b) A function that is neither one-to-one nor onto.

    (c) A function that is one-to-one but not onto.

    (d) A function that is onto but not one-to-one.

2. Prove that in a room with n people, $n \geq 2$, at least two people know exactly the same number of people. Assume knowing is a symmetric relation: If Paul knows Pat, then Pat knows Paul.

3. Infinite Acres Spa and Math Camp has an infinite number of single occupancy rooms, numbered with each positive integer. You are the night manager. The spa is fully booked for the weekend and all rooms are occupied. A bus arrives late Friday night. You find that the manager has booked an additional infinite busload of customers, with confirmation codes numbered $1, 2, 3, \ldots$. Can you accomodate the new arrivals?

4. Prove that the set of finite strings of 0's and 1's is countable.

Chapter 19

Function Composition

19.1 Reading Assignment

Read Section 7.3 of *Applied Discrete Structures*. The key concept in this section is function composition, which is fundamental in mathematics and computer science. Be sure to learn how it works and you'll be in good shape for the rest of the book!

Question 19.1.1 Response Question. Google "linux piping" and describe how this technique is related to function composition. □

Also, turn in solutions to these exercises:

Exercises

1. Define s, u, and d, all functions on the integers, by $s(n) = n^2$, $u(n) = n + 1$, and $d(n) = n - 1$. Determine:

   (a) $u \circ s \circ d$

   (b) $s \circ u \circ d$

   (c) $d \circ s \circ u$

   Describe each function with a formula similar to the way that individual functions s, u and d were defined.

2.

   - Does $f : \mathbb{Z} \to \mathbb{Z}$ defined by $f(x) = 2x + 1$ have an inverse? If it does, what is it? If it doesn't, why?

   - Does $g : \mathbb{R} \to \mathbb{R}$ defined by $g(x) = 2x + 1$ have an inverse? If it does, what is it? If it doesn't, why?

19.2 In-Class Exercises

Exercises

1. Let $A = \{1, 2, 3, 4\}$. Define $f : A \to A$ by $f(1) = 2$, $f(2) = 3$, $f(3) = 4$, and $f(4) = 1$. Find f^2, f^3, f^4 and f^{-1}. You can describe each of these functions as being equal to a previous one, or in the same manner as f was originally described.

2. Prove that if a function has an inverse, that inverse must be unique.

3. Let f and g be functions whose inverses exist. Prove that $(f \circ g)^{-1} = g^{-1} \circ f^{-1}$.

4.

   (a) Our definition of cardinality states that two sets, A and B, have the same cardinality if there exists a bijection between the two sets. Why does it not matter whether the bijection is from A into B or B into A?

   (b) Prove that "has the same cardinality as" is an equivalence relation on sets.

Chapter 20

Recursion and Sequences

20.1 Reading Assignment

Read Sections 8.1 and 8.2 of *Applied Discrete Structures*. You will see a variety of things that are most easily described through self-reference. Most computer languages allow recursion, which can be a powerful tool when used properly.

Question 20.1.1 Response Question. Recursion is used in both mathematics and computer programming. Most programming languages allow recursion and they use something called a *stack* to allow a function to "call" itself, such as in the python definition for the Binary Search Algorithm, p. 50. Google "what is a stack" and briefly describe, in your own words, what you've learned. □

Also, turn in solutions to these exercises:

Exercises

1. Consider a sequence of strings, $L(n)$ defined recursively by $L(n) = L(n-2) + L(n-2) + L(n-1)$ with $L(0) = $ "1" and $L(1) = $ "0". Here, the plus sign is taken as concatenation of strings. Determine $L(4)$.

2. Consider sequence Q defined by $Q(k) = 2k + 9$, $k \geq 1$. Complete the table below and determine a recurrence relation that describes Q.

k	$Q(k)$	$Q(k) - Q(k-1)$
2		
3		
4		
5		
6		
7		

20.2 In-Class Exercises

Exercises

1. What is computed by the following function on the natural numbers?

$$f(n) = \begin{cases} 2f(n-1) + 1 & n > 0 \\ 1 & n = 0 \end{cases}$$

2. Describe what the following function, f, does on the positive integers.

$$f(n) = \begin{cases} n & n \text{ odd} \\ 3 \cdot f(n/2) & n \text{ even} \end{cases}$$

3. I'm thinking of a number between 1 and 25. If I know you will use the binary search algorithm to guess my number and I want you to use as many guesses as possible, what are the best numbers for me to think about?

4. The length of a string is the number of characters in the string. Let $v(n)$ be the length of $L(n)$, which was defined in the the homework problems. Find a recursive description of $v(n)$.

5. *Fun Question.* This last question comes from the BBC quiz show "Round Britain". I expect that you would need to Google some of these references. If Barker and Corbett encountered, sequentially, Kieslowski's colours, Blyton's adventurers, Tarantino's undesirables and Thurber's timepieces, how many pilots would they meet next? The answer will appear at the end of the next chapter.

20.3 Binary Search Algorithm

Here is a python version of the binary search algorithm.

```python
def BinarySearch(r,j,k,C):
    found = False
    if j <= k:
        mid = floor((j + k)/2)
        print('probing at position '+str(mid))
        if r[mid] == C:
            location = mid
            found = True
            print('found in position '+str(location))
            return location
        else:
            if r[mid] > C:
                BinarySearch(r,j, mid - 1,C)
            else:
                BinarySearch(r,mid + 1,k,C)
    else:
        print('not found')
        return False
```

The output from an example of a search for the number 30 in a list of 28 numbers follows. It should be noted that in python indices start at 0, so we initially look for 30 in the entries indexed from 0 to 29. Also, probing position 13 means looking at the 14th entry in the list.

```
s=[1,9,13,16,30,31,32,33,36,37,38,45,49,50,52,61,63,64,69,77,79,80,81,83,86,90,93,96]
BinarySearch(s,0,len(s)-1,30)
```

Output:

```
probing at position 13
probing at position 6
probing at position 2
probing at position 4
found in position 4
```

Chapter 21

Solving Linear Recurrence Relations I

21.1 Reading Assignment

Read the first three subsections of Section 8.3 of *Applied Discrete Structures*. This will take you up to, but not including the section titled "Solution of Nonhomogeneous Finite Order Linear Relations". Again, the key is to make sure you become familiar with the terms that are introduced in this section.

Question 21.1.1 Response Question. One of the main reasons why recurrence relations are part of this course is that the time and/or memory needs of a computer algorithm are often measured by first identifying a recurrence relation. Once solved, many sorting algorithm are found to take a time that is proportional to n^2 to sort n items. If you are using an algorithm of this type, and it takes three minutes to sort a file with 10 million items, how long would you expect the algorithm to take to sort 20 million items? □

Also, turn in solutions to these exercises:

Exercises

1. Find a closed form expression that for the sequence $S(n)$ if $S(0) = 4$ and $S(n) = 3 \cdot S(n-1)$ if $n > 0$.

2. Find a closed form expression that for the sequence $T(n)$ if $T(0) = 1$, $T(1) = 5$ and $T(n) - 3 \cdot T(n-1) - 4 \cdot T(n-2) = 0$ if $n > 2$.

21.2 In-Class Exercises

Exercises

1. Find a closed form expression that for the sequence $V(n)$ if $V(0) = 2$, $V(1) = 3$ and $V(n) = \frac{1}{2} \cdot V(n-1) + \frac{1}{2} \cdot V(n-2)$ if $n \geq 2$.

2. Find a closed form expression that for the sequence $Q(n)$ if $Q(0) = 3$, $Q(1) = 0$ and $Q(n) = 6 \cdot Q(n-1) - 9 \cdot Q(n-2)$ if $n \geq 2$.

3. The recurrence relation $R(n) = R(n-1)+2^n$, $n \geq 1$ is non-homogeneous. This is the subject of the next class, but it can be turned into a second order homogeneous recurrence relation. This can be done by replacing n with $n-1$ in the recurrence relation and multiplying that equation by 2. You can then eliminate the 2^n term. Find the general solution to the resulting second order recurrence relation.

4. The Fibonacci Sequence, p. 54 is a second order homogeneous linear recurrence relation. It's characteristic roots are no so nice and clean as some of the examples we've seen, but developing a closed form solution is made easier by the fact that if α and β are its two characteristic roots, then $\alpha + \beta = 1$ and $\alpha \cdot \beta = -1$. Verify this and then solve for a closed form expression for F_k.

21.3 Fibonacci Sequence

The Fibonacci Sequence is the sequence F defined by

$$F_0 = 1, \ F_1 = 1 \text{ and}$$

$$F_k = F_{k-2} + F_{k-1} \text{ for } k \geq 2$$

Note 21.3.1 Some people prefer to start the Fibonacci numbers with 0 and 1 instead of 1 and 1. that doesn't really change the main properties of the sequence, but indices may need adjusting.

Answer to the last in-class question in the previous chapter: These clues give you numbers in ascending order: The TWO Ronnies, the THREE colours trilogy, the Famous FIVE, the Hateful EIGHT, Thurber's THIRTEEN clocks. This is a Fibonacci sequence, in which each number is the sum of the previous two. The next number in the sequence (and the answer to the question) must therefore be 21 - as in the rock band, 21 Pilots. This was a question posed at the end of Programme 7 of the 2020 Round Britain Quiz.

Chapter 22

Solving Linear Recurrence Relations II

22.1 Reading Assignment

Read the remainder of Section 8.3 starting with Subsection 8.3.4: "Solution of Nonhomogeneous Finite Order Linear Relations." Once you've completed this section - it's a long one - you should be able to solve many recurrence relations that appear in mathematics and computer science. Not all of them though!

Question 22.1.1 **Response Question.** An algorithm that sorts files in "$n \log n$-time" is normally considered better than one that sorts in "n^2-time". However, that's not always the case for smaller files. The time it takes to sort n items using Algorithms A and B take $1200 \cdot n \log_2 n$ nanoseconds and $5n^2$ nanoseconds. respectively. How large must a file be to make Algorithm A the preferred one? □

Also, turn in solutions to these exercises:

Exercises

1. Find a closed form solution to $S(k) - 2S(k-1) = 5^k$, with $S(0) = 3$

2. What form would a particular solution to $T(n) - 5 \cdot T(n-1) + 6 \cdot T(n-2) = 7 \cdot 3^n$ take? Find just a particular solution at this time.

22.2 In-Class Exercises

Exercises

1. Suppose that a computer algorithm takes no time to sort a list with one item, but if it is given a list with n items, $n \geq 2$, then it takes $T(n) = T(n-1) + 3 \cdot n$ nanoseconds. Find a closed form expression for $T(n)$

2. Find a closed form solution to $S(k) - 5S(k-1) + 6S(k-2) = 2$, with $S(0) = -1$, and $S(1) = 0$.

3. Find a closed form solution to $S(k) - 5S(k-1) + 6S(k-2) = 7 \cdot 3^k$, with $S(0) = 1$, and $S(1) = 3$.

4. If you were to deposit a certain amount of money at the end of each year for a number of years, this sequence of payments would be called an *annuity*. With an annual interest rate of 5 percent, how much would you need to deposit into an annuity to have a value of one million dollars after 18 years?

22.3 Interest

Interest is earned on investments by adding to the invested amount, called the *principle*. An interest rate is a percentage of the principle that is earned. For example if you invest $2,000 in an investment that earns 3 percent, your interest in one year would be $2,000 \cdot 0.03 = \$60$. This is added to principle. The new principle is normally computed in one step by multiplying by 1.03.

$$\$2,000 + \$2,000 \cdot 0.03 = \$2,000 \cdot 1.03 = \$2,060.$$

Chapter 23

Some Common Recurrence Relations

23.1 Reading Assignment

Read Section 8.4 of *Applied Discrete Structures*. There is no general method for solving all recurrence relations. In this section we consider a few cases for which the method in Section 8.3 cannot be applied.

Question 23.1.1 Response Question. In this section we study algorithms for searching and sorting. If you have data that isn't sorted, then the binary search algorithm can't be implemented and you must do a sequential search. In a sequential search your look at each item in a list until you find what you're looking for, or you reach the end of the list. What is the average number of items you will examine in a successful, and in an unsuccessful search of a list with n items? □

Also, turn in solutions to these exercises:

Exercises

1. Prove that if $n \geq 0$, $\lfloor n/2 \rfloor + \lceil n/2 \rceil = n$.
2. One derangement of $\{1, 2, 3, 4\}$ is 2143. List all others.

23.2 In-Class Exercises

Exercises

1. The *selection sort* algorithm on a list of n proceeds first by finding the largest item in the list and placing it last, exchanging it with the n-th item, if necessary. Then a selection sort sort of the first $n - 1$ items is conducted. Let $C(n)$ be the number of comparisons needed to complete a selection sort of n items. Find a recurrence relation and initial condition for C and solve it.

2. Suppose $n \geq 2$ and $1 \leq k \leq n$. How many permutations of $\{1, 2, \ldots, n\}$, have the property that k is a fixed point? The set of all such permutations is called U_k in the next problem.

3. Count the number of derangements of $\{1, 2, 3, 4\}$ using inclusion-exclusion, p. 58. Do this by counting the non-derangements in the union $U_1 \cup U_2 \cup U_3 \cup U_4$, where U_k is the set of permutations for which k is fixed. You can subtract that result from 4! Generalize to an arbitrary value of n.

4. Among all continuous functions on the interval $[0, 1]$, how many are derangements in that they have no fixed points?

23.3 Inclusion-Exclusion

Here are the two and three set Inclusion-Exclusion Laws. You'll need to generalize to four sets and later to n sets in 3, p. 58, but all of the sets are similar so it isn't as complicated a you might think.

Theorem 23.3.1 Laws of Inclusion-Exclusion. *Given finite sets* $A_1, A_2, A_3,$ *then*

(a) *The Two Set Inclusion-Exclusion Law:*

$$|A_1 \cup A_2| = |A_1| + |A_2| - |A_1 \cap A_2|$$

(b) *The Three Set Inclusion-Exclusion Law:*

$$\begin{aligned}
|A_1 \cup A_2 \cup A_3| = &|A_1| + |A_2| + |A_3| \\
&- (|A_1 \cap A_2| + |A_1 \cap A_3| + |A_2 \cap A_3|) \\
&+ |A_1 \cap A_2 \cap A_3|
\end{aligned}$$

Chapter 24

Generating Functions

24.1 Reading Assignment

Read the first two subsections of Section 8.5 of *Applied Discrete Structures*. Generating functions can serve as an alternative to the algorithm we introduced in Section 8.3, but also is a tool for solving other problems.

Question 24.1.1 Response Question. (adopteds from [1]) Suppose we treat addition as logical *or* and multiplication as logical *and*. Furthermore, suppose that B stands for banana. Then to say you could have as many as two bananas, we could write $B^0 + B^1 + B^2 = 1 + B + B^2$. Suppose that in addition you could have up to three apples (use A for apples) and zero or one pears (use P for pears). What algebraic expression represents all your choices in selecting fruits? Identify the part of this expression where you select exactly two pieces of fruit.   □

Also, turn in solutions to these exercises:

Exercises

1. What sequence has as its generating function $\frac{1}{3-2x}$?

2. How are the generating functions of the sequences $S(n) = n^2$ and $T(n) = (n+1)^2$ related?

24.2 In-Class Exercises

Exercises

1. Let
$$d(n) = \begin{cases} 1 & \text{if } 1 \leq n \leq 6 \\ 0 & \text{if } n = 0 \text{ or } n > 6 \end{cases}.$$

    What sequence has as its generating function $G(d; z)^2$? How is that sequence related to what you get when you roll two dice and add the top faces?

2.   Earlier, we proved that with supplies of five and eight cent stamps, we could make any postage amount of 28 cents or more. Here, we will look at what smaller amounts can and can't be made. Let $F(z) = \sum_{n=0}^{\infty}(z^5)^n$ and $E(z) = \sum_{n=0}^{\infty}(z^8)^n$. Every combination of stamps corresponds with the product of one term from $F(z)$ with one term from $E(z)$. For example, the product $(z^5)^2 \cdot (z^8)^1 = z^{18}$ corresponds with combining two five cent stamps and one eight cent stamp. Compute the first few terms of $F(z) \cdot E(z)$ to get all terms with degree less than 28. The terms that are missing (have a coefficient of zero) are the ones that correspond with amounts that can't be created. In general, the coefficient of z^n in the product will be the number of ways that n cents can be made. Do a similar calculation to identify the amounts that cannot created with 7 and 9 cent stamps.

3.   How many ways can you give someone fifty cents using any number of nickels, dimes, and quarters?

Part II

Structures

Chapter 25

Start of Second Semeser, Review

In the first class of the second semester, some time will be taken to review the course format and syllabus. In the remaining time, work on problems that review some of the fundamental concepts from the first semester.

25.1 In-Class Exercises

Exercises

1. Prove that if f and g are bijections on a set X, then $g \circ f$ is also a bijection on X.

2. Partition the set $X = \{k \in \mathbb{Z} \mid -10 \leq k \leq 10\}$ into equivalence classes according to the relation congruence modulo 4, $\equiv_4$.

3. Is the following logical argument valid?

    Peacham is either in Vermont or New Hampshire. If Peacham is in Vermont, then Peacham is in New England. If Peacham is in New Hamphire, then Peacham is in New England. Therefore, Peacham is in New England.

4. Prove, by mathematical induction, that $F_0 + F_1 + F_2 + \cdots + F_n = F_{n+2} - 1$, where F_n is the nth number in the Fibonacci Sequence, p. 54.

5. Prove that if n is an integer and you divide n^2 by 5, then the remainder is always 0, 1, or 4.

6. Prove that the square root of 5 is an irrational number.

Chapter 26

Graphs

26.1 Reading Assignment

Read Section 9.1 of *Applied Discrete Structures*. This section introduces some basic terminology and properties of graphs.

Question 26.1.1 Response Question. The original version of the computer language BASIC on Apple II computers used 16 bits for an integer, one of which was used to specify the sign of the integer. What would have been the maximum size of an integer in that language? ☐

Also, turn in solutions to these exercises:

- Why is the sum of the degrees of the vertices of any undirected graph always even?

- Demonstrate that $(4, 3, 2, 2, 1)$ is a graphic sequence.

26.2 In-Class Exercises

1. Prove that any graph with at least two vertices must have two vertices of the same degree.

2. Starting at vertex s, any finite path in Figure 26.2.1, p. 66 produces a string of bits that are recorded according to the labels on each edge. For example, the path s, a, b, a, b, b, a will produce the string 101001.

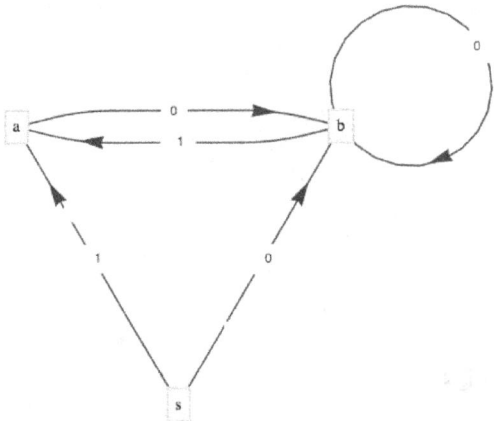

Figure 26.2.1 A model for bit strings with no consecutive 1's

This graph produces bit strings that contain no consecutive 1's. Draw a graph similar to it that produces bit strings containing no more than two consecutive 1's.

3. Find two isomorphisms between the following graphs.

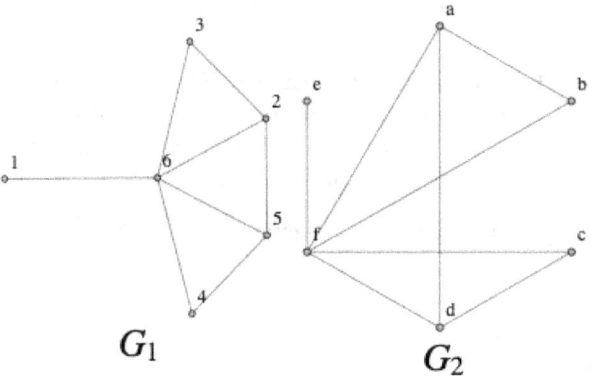

Figure 26.2.2 Two Isomorphic Graphs

4. (a) Determine whether the following sequences are graphic. Explain your logic if the answer is "no" and draw a graph with the sequence as its degree sequence if it is "yes."

    i $(5, 1, 1, 1, 1, 1)$
    ii $(3, 3, 3, 3)$
    iii $(3, 3, 3, 3, 3)$
    iv $(4, 3, 2, 1, 0)$
    v $(2, 2, 2, 1, 1)$
    vi $(3, 2, 2, 2, 1)$

  (b) Based on observations you might have made in from the examples above, describe as many characteristics as you can about graphic sequences.

Chapter 27

Data Structures and Connectivity

27.1 Reading Assignment

Read Sections 9.2 and 9.3 of *Applied Discrete Structures*. The topics in these sections are data structures for graphs and connectivity between vertices in a graph.

Question 27.1.1 Response Question. You're certainly more familiar with the terms diameter, radius and center as they pertain to circles. Compare their use for circles with their use in graph theory. How are they similar and how are they different? □

Also, turn in solutions to these exercises:

Exercises

1. Draw the undirected graph that is represented by the following Sage-Math expression. `Graph({1:[2,3,4],2:[3,5],3:[5,7],4:[6,7]})`

2. In a breadth first search starting at vertex 2 of the graph in the previous problem, what would be the depth sets?

27.2 In-Class Exercises

Exercises

1. Determine the eccentricities of each vertex, and the diameter, radius, and center of the graph in exercise 1 of the reading assignment.

2. Determine the diameter, radius, and center of the graph with the following distance matrix, where $D_{i,j}$ is the length of the shortest

path from i to j.

$$D = \begin{pmatrix} 0 & 2 & 2 & 2 & 3 & 1 & 1 & 3 & 3 & 1 & 2 & 2 \\ 2 & 0 & 3 & 3 & 2 & 2 & 1 & 4 & 1 & 2 & 2 & 1 \\ 2 & 3 & 0 & 2 & 5 & 3 & 2 & 3 & 4 & 1 & 4 & 3 \\ 2 & 3 & 2 & 0 & 3 & 1 & 2 & 1 & 3 & 1 & 2 & 3 \\ 3 & 2 & 5 & 3 & 0 & 2 & 3 & 4 & 1 & 4 & 1 & 3 \\ 1 & 2 & 3 & 1 & 2 & 0 & 1 & 2 & 2 & 2 & 1 & 2 \\ 1 & 1 & 2 & 2 & 3 & 1 & 0 & 3 & 2 & 1 & 2 & 1 \\ 3 & 4 & 3 & 1 & 4 & 2 & 3 & 0 & 4 & 2 & 3 & 4 \\ 3 & 1 & 4 & 3 & 1 & 2 & 2 & 4 & 0 & 3 & 1 & 2 \\ 1 & 2 & 1 & 1 & 4 & 2 & 1 & 2 & 3 & 0 & 3 & 2 \\ 2 & 2 & 4 & 2 & 1 & 1 & 2 & 3 & 1 & 3 & 0 & 3 \\ 2 & 1 & 3 & 3 & 3 & 2 & 1 & 4 & 2 & 2 & 3 & 0 \end{pmatrix}$$

Can you construct the graph from this matrix?

3. A classroom has 5 rows of desks, with 7 desks in each row. Suppose we want to represent this rectangular arrangement of desks in an undirected graph, where each desk is connected to the neighboring desks to it's front, back, left and right. Of course, some desks have fewer than four neighbors.

(a) How many edges will the graph have?

(b) Determine the possible eccentricities of vertices in this graph. What is the diameter, radius and center of the graph.

(c) If an adjacency matrix is constructed for this graph with one bit (0/1) for each entry, how many bits would be needed? Assume we only store the part of the matrix that appears to the top right of the main diagonal of the matrix since its symmetric.

(d) If an edge dictionary is used in which eight bits are used for each vertex number and eight bits for each entry in the list of adjacent vertices, how many bits would be needed? Assume that if i appears in the list of neighbors of j, then we don't need to list j in the list of neighbors of i.

(e) If a list of ordered pairs is used, where each pair requires 16 bit, how many bits would be needed? Assume that for any two vertices, i and j, only one of the pairs (i, j) and (j, i) need to be listed.

Chapter 28

Graph Traversals

28.1 Reading Assignment

Read Section 9.4 of *Applied Discrete Structures*. In this section, we consider whether certain paths or circuits exist in a given graph. Our objective will either be to trace every edge once or visit every vertex once.

Question 28.1.1 Response Question. In the 18th century, Koenigsberg was part of Prussia. In what country is it now? Besides its bridges, find one other fact about Koenigsberg. □

Also, turn in solutions to these exercises:

Exercises

1. In answering the question "Is every Eulerian graph also Hamiltonian?" Hansel pointed out an Eulerian circuit in a specific graph that visited several vertices more than once. He concluded that this graph can't be Hamiltonian since a Hamiltonian circuit visits each vertex exactly once. Therefore, the answer to the question is "No!" Is his answer correct? Is his reasoning correct? Explain your answers.

2. How many different Hamiltonian circuits are there of a K_n, $k \geq 3$, that start with the edge $(1, 2)$?

28.2 In-Class Exercises

Exercises

1. For each of the following sets of twelve dominos, is it possible to arrange them end-to-end so that the numbers every two touching ends have matching numbers?

(a)

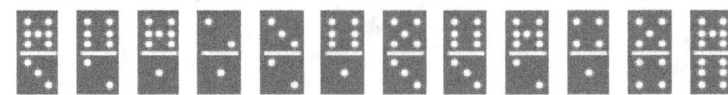

Figure 28.2.1

(b)

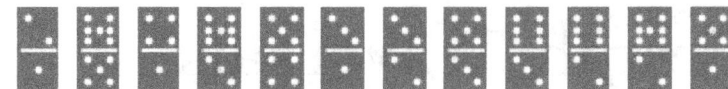

Figure 28.2.2

2. A telephone company employee needs to check the telephone lines hanging from telephone poles for a cut in the line over a grid of streets in a city without service. What graph theory problem is she going to have to solve if she wants to do this efficiently?

3. King Arthur and his knights are either friends with one another or are enemies. He has called a meeting of all of his knights and would like to have them seated at his huge round table so that each knight is seated next to two friends. What graph theory problem is he going to have to solve?

4. The Gray code for the 3-cube is not unique in that there are other Hamiltonian circuits that are not equal to G_3 or its reverse. Find one. Is it your different in "shape" from G? For the n-cube, $n > 3$, are there Hamiltonian paths that are different in shape from G_n?

5. Assume we number positions in the Gray Code G_n from 0 to $2^n - 1$. What is the bit string of in position 20 in the Gray code for the 6-cube? Where will you find 000101 in G_6? In what position in G_{10} will you find 0010010000?

Chapter 29

Planarity and Graph Coloring

29.1 Reading Assignment

Read Section 9.6 of *Applied Discrete Structures*. There are two main topics of this section. The first is planarity of graphs, whether a graph can be drawn on paper without having any edges crossing one another. The second is coloring, the vertices of a graph are assigned colors with the rule that two vertices that are connected by an edge must have different colors. The two topics are unified by the Four Color Theorem.

Question 29.1.1 Response Question. Who was István Fáry and what did he prove about planar graphs? ☐

Also, turn in solutions to these exercises:

Exercises

1. Although a $K_{3,3}$ it is not planar, it can be embedded on a torus without any edge crossings. This is demonstrated in the video https://www.youtube.com/watch?v=k2O0Av_8_fo. Watch the video and then dermine whether you can embed a K_5 on a torus without any edge crossings.

2. Find a coloring of the following graph with as few colors as possible. Use the letters $A, B, C, \ldots >$ for colors and submit your answer in the form of a partition of $\{1, 2, \ldots, 8\}$:

   $\{$set of vertices with color A, set of vertices with color $B, \ldots\}$

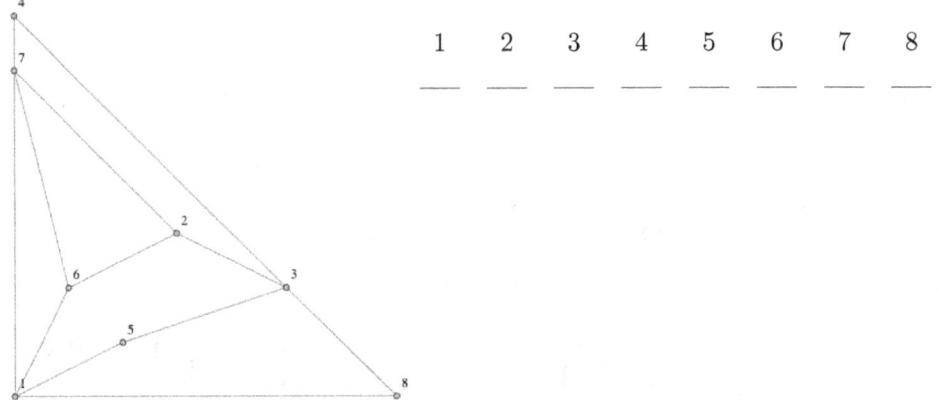

Figure 29.1.2

29.2 In-Class Exercises

Exercises

1. If a planar graph G with 12 vertices divides the plane into 21 regions, how many edges must G have?

2. The graph G has vertices with degree sequence $(5, 3, 3, 2, 2, 1)$. How many edges does it have? If it is planar, into how many regions of the plane does it divide? Does G have an Eulerian path?

3. Suppose an undirected planar graph G has the following properties:

    - All vertices have degree three (this called a **cubic graph**)

    - All faces of the graph's planar embedding are hexagons and pentagons; i. e, they is made up of either five or six edges.

    What can be said about the numbers of hexagons and pentagons in the graph?

4. A teacher decides to have their students grade one another's quizzes. The scheme that is devised is that when a bell is rung, each student is to immediately pass their quiz to a classmate either to the front, back, right or left of themselves. Suppose that class is arranged in rectanglar configuration of five rows of seven students each. Is is possible for each student to get exactly one quiz to grade?

5. The chromatic polynomial of a graph, G is a polynomial $g(x)$ such that for each positive integer n, $g(n)$ equals the number of different ways you can color G with n colors. For example, the chromatic polynomial of a K_3 is $x(x-1)(x-2)$.

    (a) What is the chromatic polynomial of a K_4?

    (b) What is the chromatic polynomial of a C_4, a cycle with four vertices?

    (c) What is the chromatic polynomial of a $K_{3,3}$?

    (d) How are the chromatic number and the chromatic polynomial of a graph related?

Chapter 30

Trees and Spanning Trees

30.1 Reading Assignment

Read Sections 10.1 and 10.2 of *Applied Discrete Structures*. In these sections we introduce trees, which are connected graphs without cycles. We also examine spanning trees of connected graphs, which are the smallest subgraphs which maintain connectedness in a graph.

Question 30.1.1 Response Question. In general, a family tree isn't really a tree in the sense of graph theory. Explain why. You can assume links in the family tree are undirected.   □

Also, turn in solutions to these exercises:

Exercises

1. Given a tree with n vertices, $n \geq 2$, how many leaves (vertices of degree 1) could it have?

2. What can you say about the sum of the entries in the degree sequence of a tree?

30.2 In-Class Exercises

Exercises

1. Use the planarity of trees together with Euler's Formula, p. 76 to derive the relationship between the number of vertices and number of edges in a tree.

2. Prove that every tree is bipartite, i. e., has a 2-coloring.

3. Prove that every connected graph which is not itself a tree must have at last three different spanning trees.

4. Use Prim's algorithm starting at vertex 1 to find a minimal spanning tree for the following graph.

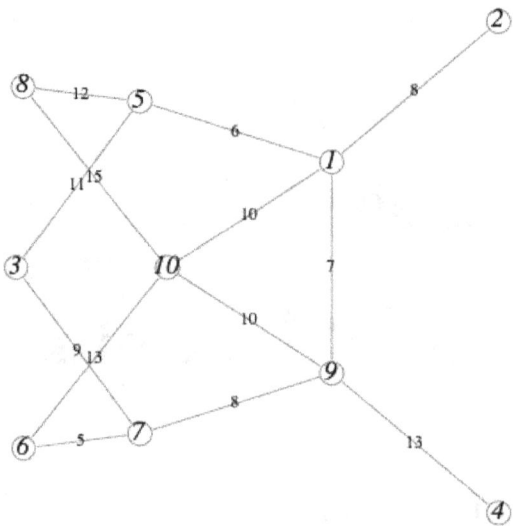

Figure 30.2.1

5. For each degree sequence below, decide whether it must always, must never, or could possibly be a degree sequence for a tree. If it is always a tree, is the tree unique? Justify your answers. (Adapted from [3].)

   (a) $(3, 3, 2, 2, 2)$

   (b) $(3, 2, 2, 1, 1, 1)$

   (c) $(3, 3, 3, 1, 1, 1)$

   (d) $(4, 4, 1, 1, 1, 1, 1, 1)$

30.3 Euler's Formula

Theorem 30.3.1 Euler's Formula. *If $G = (V, E)$ is a connected planar graph with r regions, v vertices, and e edges, then*

$$v + r - e = 2$$

Chapter 31

Rooted Trees, Binary Trees

31.1 Reading Assignment

Read Sections 10.3 and 10.4 of *Applied Discrete Structures*. Here we consider some more structured trees in which a specific vertex is designated as the *root* of the tree. We see that there are several applications of these trees. Special attention is given to binary trees, where no more than two edges branch away from the root and any away other vertex that is reached by the root.

Question 31.1.1 Response Question. The level of a vertex of a binary tree is the length of the path from the root to that vertex. What is the maximum number of vertices at levels $1, 2, 3, \ldots$ of any binary tree? □

Also, turn in solutions to these exercises:

Exercises

1. Draw all different ordered rooted trees with four vertices.

2. Construct the expression trees for the expressions $\frac{a \cdot x + b}{c \cdot x + d}$ and and $\frac{a}{c} + \frac{r}{c \cdot x + d}$.

31.2 In-Class Exercises

Exercises

1. Given the list of data $[483, 569, 150, 649, 659, 198, 356, 241, 359, 930]$, insert the data sequentially into a binary sorting tree.

2. List the vertices in a postorder traversal of the expression tree of $\frac{ax+b}{cx+d}$.

3. Suppose $0 < k < n$. Define the $\binom{n}{k}$-tree to be the binary tree with $\binom{n}{k}$ at its root and left and right subtrees having roots $\binom{n-1}{k-1}$ and $\binom{n-1}{k}$, respectively. If $k = 0$ or $k = n$, then the $\binom{n}{k}$-tree is a leaf containing the number 1.

    (a) Draw the $\binom{5}{1}$ and $\binom{5}{2}$ trees.

(b) How many leaves does a $\binom{n}{k}$-tree contain? Prove your answer.

4. How many binary trees are there having n vertices and an empty right subtree?

    Hint. Use the formula for the number of binary trees with n vertices.

5.

There is a one to one correspondence between ordered rooted trees and binary trees. If you start with an ordered rooted tree, T, you can build a binary tree B with an empty right subtree by placing the the root of T at the root of B Then for every vertex v from T that has been placed in B, place it's left-most child (if there is one) as v's left child in B. Make v's next sibling (if there is one) in T the right child in B.

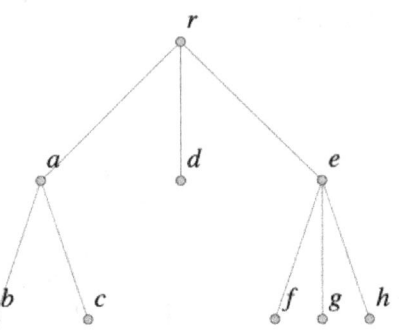

Figure 31.2.1 An ordered rooted tree with root r.

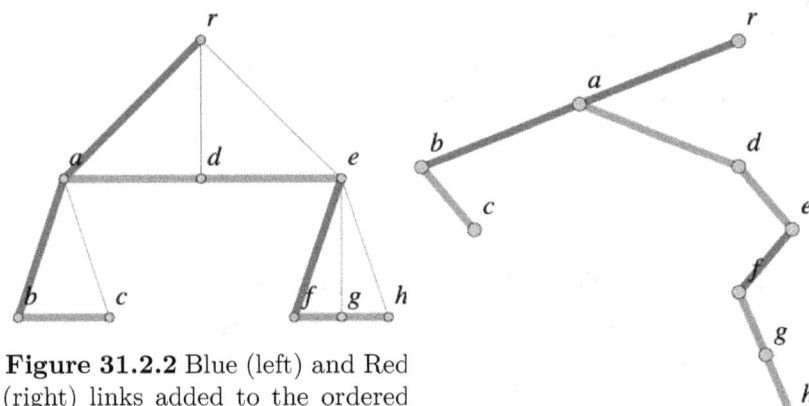

Figure 31.2.2 Blue (left) and Red (right) links added to the ordered rooted tree with root r.

Figure 31.2.3 Binary tree corresponding to the ordered rooted tree.

(a) Why will B have no right children in this correspondence?

(b) Draw the binary tree that is produced by the ordered rooted tree in Figure 31.2.4, p. 79.

(c) Draw the ordered tree that produces the binary tree in Figure 31.2.5, p. 79.

(d) The left subtree of the binary tree in Figure 31.2.5, p. 79 is one of 5 different binary trees with three vertices. Draw each of them and also the ordered rooted tree that each corresponds with.

(e) What does this correspondence tell us about how the numbers of different binary trees and ordered rooted trees are related?

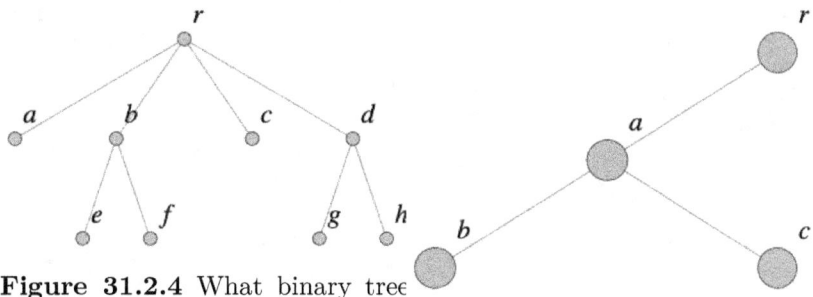

Figure 31.2.4 What binary tree does this correspond with?

Figure 31.2.5 What ordered rooted tree does this correspond with?

Chapter 32

Binary Operations

32.1 Reading Assignment

Read Section 11.1 of *Applied Discrete Structures*. This is the beginning of a new topic, but you'll find that there are connections between algebra and graph theory. In this first section we review properties of binary operations.

Question 32.1.1 Response Question. True or False? Multiplication is an associative operation on the set $\{-1, 1\}$. Justify your answer. □

Also, turn in solutions to these exercises:

Exercises

1. Explain why subtraction is not an associative operation on the integers.

2. Write out the operation table for intersection on the $\mathcal{P}(\{0, 1\})$, the power set of $\{0, 1\}$.

32.2 In-Class Exercises

Exercises

1. What properties that are introduced in Secton 11.1 are true of intersection on the power set of $\{0, 1\}$?.

2. Consider the operation □ on the integers defined by

$$a \square b = a + b - 1.$$

   (a) Is □ associative?

   (b) Does □ have an identity in the positive integers?

   (c) Does □ have the inverse property?

3.   This question is open-ended, with no specific "right" answer. Suppose
that V is some finite nonempty set. What would be an example of a
binary operation on the set of all undirected graphs with vertex set V?

Chapter 33

Algebraic Structures

33.1 Reading Assignment

Read Section 10.2 of *Applied Discrete Structures*. We start by introducing the idea of levels of abstraction as they apply to monoids. We then introduce groups, which is the main type of system we focus on in this chapter.

Question 33.1.1 Response Question. Discuss the connection between groups and monoids. Is every monoid a group? Is every group a monoid? □

Also, turn in solutions to these exercises:

Exercises

1. Write out the operation table for multiplication on the set $\{1, -1, i, -i\}$.

2. How many different commutative binary operations are there on any two element set?

33.2 In-Class Exercises

Exercises

1. Let U be a any set. Prove that, $\oplus$, defined by $A \oplus B = (A \cup B) - (A \cap B)$ is an associative operation on $\mathcal{P}(U)$.

2. Let U be a any set. Prove that, $[\mathcal{P}(U); \oplus]$ is a group.

3. Consider the following set of six algebraic expressions, each defining a function on the set of real numbers excluding the numbers 0 and 1.

$$\mathcal{H} = \left\{ x, 1 - x, \frac{1}{1-x}, \frac{1}{x}, \frac{x-1}{x}, \frac{x}{x-1} \right\} = \{y_1, y_2, y_3, y_4, y_5, y_6\}$$

We can opperate on any two of these expressions using function composition. For example,

$$(y_3 \circ y_4)(x) = y_3(y_4(x)) = y_3\left(\frac{1}{x}\right) = \frac{1}{1-\frac{1}{x}} = \frac{x}{x-1} = y_6(x)$$

Therefore, $y_3 \circ y_4 = y_6$. Complete the following operation table for function composition on $\mathcal{H}$.

$\circ$	y_1	y_2	y_3	y_4	y_5	y_6
y_1	y_1	y_2	y_3	y_4	y_5	y_6
y_2	y_2	y_1	y_6	y_5	y_4	
y_3	y_3	y_4		y_6		
y_4		y_3	y_2			
y_5						
y_6						

Figure 33.2.1 Partially completed operation table for $\mathcal{H}$

Is $[\mathcal{H}, \circ]$ a monoid? Is it a group?

4. Let S be the set of all strings of letters of length zero or more from the set $\{a, b, c\}$ that do not have any consecutive identical letters. Define $*$ on S by the rule that if $x, y \in S$ then $x * y$ is determined by first concatenating x and y and then repeatedly removing any occurrences of two identical letters until the result is an element of S. Is $[S, *]$ a monoid? Is it a group?

Chapter 34

Properties of Groups

34.1 Reading Assignment

Read Section 11.3 of *Applied Discrete Structures*. In this section we examine some of the basic theorems that follow immediately from the axioms of a group. Reviewing the proofs is a good exercise in solidifying an understanding of the axioms. In most cases, these theorems are applied in later work without explicitly stating them.

Question 34.1.1 Response Question. SageMath can be used to explore concrete groups. It can't be used to prove significant theorems, but you can verify that the theorems are true. Here is one particular group's operation table generated using Sage. The group, with domain [a,b,c,d,e,f,g,h], is in a family called the dihedral groups, and will be encountered in Section 15.3. For the purposes of this question, notice that in the body of the table, each element appears exactly once in each row and column. Which theorem guarantees that this is always true for a group's operation table?

```
G=DihedralGroup(4)
G.cayley_table()
```

```
*  a b c d e f g h
+----------------
a| a b c d e f g h
b| b a d c f e h g
c| c g a e d h b f
d| d h b f c g a e
e| e f g h a b c d
f| f e h g b a d c
g| g c e a h d f b
h| h d f b g c e a
```

Answer. Any of of the following theorems could be applied to make this observation: Cancellation Laws, Linear Equations in a Groups, or the Pigeonhole Principle. □

Also, do these exercises:

Exercises

1.   Suppose $[G; *]$ is a group with $a, b, c \in G$. Solve the equation

$$a * x * a^{-1} * b = c$$

for x.

2.   Notice that $\{1, 2, 3\} \cap \{1, 2\} = \{1, 2, 3, 4\} \cap \{1, 2\}$, yet $\{1, 2, 3\} \neq \{1, 2, 3, 4\}$. Does this contradict the Cancellation Laws, p. 86? Explain your answer.

34.2 In-Class Exercises

Exercises

1.   Prove that if $a^2 = e$ for all elements a in a group G, then G must be abelian.

    Hint.   Given $a, b \in G$, apply the premise to $a * b$.

2.   Suppose $[G; *]$ is a group and $a \in G$. Define $f_a : G \to G$ by $f_a(x) = a * x$. If we compose two such functions, f_a and f_b where $a, b \in G$, what function do we get for $f_a \circ f_b$?

3.   This problem anticipates a future topic that you can plausibly discover on your own. Functions like the ones in the previous problem can be of use in this one.

    Suppose $[G; *]$ is a finite group of order n. The last theorem in Section 11.3, p. 86 in Section 10.3 states that if $a \in G$, there exists a positive integer, m less than n such that $a^m = e$. Prove that the least such positive integer is a divisor of n.

    Hint.   You can partition G into subsets of equal cardinality.

34.3 Some Theorems

Two of the theorems in Section 10.3 follow. In doing the exercises, you don't need to prove these theorems or the others in the section.

Theorem 34.3.1   Cancellation Laws. *If a, b, and c are elements of group G, then*

left cancellation: $(a * b = a * c) \Rightarrow b = c$
right cancellation: $(b * a = c * a) \Rightarrow b = c$

Theorem 34.3.2   last theorem in Section 11.3. *If G is a finite group, $|G| = n$, and a is an element of G, then there exists a positive integer m such that $a^m = e$ and $m \leq n$.*

Chapter 35

Modular Arithmetic

35.1 Reading Assignment

Read Section 11.4 of *Applied Discrete Structures*. The definitions of modular addition and multiplication in this section are *really important*. They are used throughout the rest of the book. The section starts with a review of division properties for integers in order to set up the modular operations. Generally, the mod n operations are operations only on either $\mathbb{Z}_n$, or a subset of $\mathbb{Z}_n$. For example, it wouldn't make sense to do mod 10 addition on Z_{12}.

Question 35.1.1 Response Question. If the product of two numbers is zero, must one of the numbers equal zero? □

Also, turn in solutions to the following exercises. Both of them could be done using SageMath by typing two short inputs, but you are encouraged to write (or type) both of these be hand.

Exercises

1. Compute by hand the greatest common divisor of 2028 and 1001. Show the calculations you did to get your answer.

2. Write out the operation table for mod 4 addition on $\mathbb{Z}_4$. Identify the inverse of each element.

35.2 In-Class Exercises

Exercises

1. Let n be a positive integer greater than 1. Show that $n - 1$ has a multiplicative inverse mod n.

2. Compute by hand the additive and multiplicative inverses of 653 mod 1001. Show the calculations you did to get your answer.

3. Write out the operation table for the multiplicative group of integers mod 8 on $\mathbb{U}_8 = \{k \in \mathbb{Z}_8 \mid \gcd(8, k) = 1\}$. Identify the inverse of each element.

4. Write out the operation table for mod 10 multiplication on $T = \{0, 2, 4, 6, 8\}$. Is $[T; \times_{10}]$ a monoid? Is it a group?

5. Find a formula for the inverse of the function $g : \mathbb{Z}_{29} \to \mathbb{Z}_{29}$ where $g(a) = 5 \times_{29} a +_{29} 25$. Suggestion: You can use the symbols $\times$ and $+$ in your work instead of $\times_{29}$ and $+_{29}$ as long as we agree that they stand for mod 29 operations.

Chapter 36

Subsystems

36.1 Reading Assignment

Read Section 11.5 of *Applied Discrete Structures*. The main idea is that there are certain subsets of groups that qualify as groups in their own right. This is an instance of a more general situation in mathematics where a structure can contain smaller structures. The idea of a subset is the most simple case of this, but we have also seen subgraphs in the study of graphs and trees. Finding all subgroups of a group isn't always easy, but describing and finding the cyclic subgroups of a group is not too difficult. Pay close attention how this is done!

Question 36.1.1 Response Question. Let $m\mathbb{Z} = \{m \cdot k \mid k \in \mathbb{Z}\}$. Chris says that $7\mathbb{Z}$ is a subgroup of $14\mathbb{Z}$ because $7 \leq 14$. How would you respond to Chris? $\square$

Also, turn in solutions to these exercises:

Exercises

1. True or false?: The set of all bit strings that begin with a 1 form a submonoid of the monoid of all bit strings with concatination.

2. True or false?: $\{0, 25, 50, 75\}$ is a subgroup of the additive group $\mathbb{Z}_{100}$. Explain your answer.

36.2 In-Class Exercises

Exercises

1. Determine the cyclic subgroup generated by 4 in the group $\mathbb{Z}_{11}$
2. Determine the cyclic subgroup generated by 4 in the group $\mathbb{U}_{11}$
3. In this exercise, we will consider the orders of the additive groups of integers modulo n for three different values of n, and then speculate on some general results.

   (a) Determine the orders of different elements of the group $\mathbb{Z}_5$.

   (b) Determine the orders of different elements of the group $\mathbb{Z}_{15}$.

(c) Determine the orders of different elements of the group $\mathbb{Z}_{16}$.

(d) Comparing the results you got above, hypothesize on what you would expect for different moduli

4. Assume that $[G; *]$ is a group and H is a nonempty subset of G. Suppose that we are told that for $a, b \in H$, we are guaranteed that $a^{-1} * b \in H$. Prove that H is a subgroup of G.

    Hint. Prove the identity, inverses and closure properties of a subgroup, in that order.

36.3 Handouts

Here is a reminder of the definition of the associative property.

Definition 36.3.1 Associative Property. Let $*$ be a binary operation on a set S. We say that $*$ is **associative** if and only if $(a * b) * c = a * (b * c)$ for all $a, b, c \in S$. $\diamond$

Chapter 37

Direct Products

37.1 Reading Assignment

Read Section 11.6 of *Applied Discrete Structures*. At some point, you may have seen an "addition" of pairs that looked something like this:

$$(2,9) + (1,-3) = (2+1, 9+(-3)) = (3,6).$$

This is an example of "coordinate-wise addition," where the first coordinate of the sum depends only on the two first coordinates of the terms on the left. This is exactly the way direct products work in general. In this section you will see what is produced from direct products of groups.

Question 37.1.1 Response Question. A fraction can be though of as an ordered pair, a/b, where a is the numerator and nonzero b is the denominator. Is addition of fractions a coordinate-wise operaton? □

Also, turn in solutions to this exercise:

Exercises

1. Write out the operation table for the group $\mathbb{Z}_2 \times \mathbb{Z}_2$

2. Find all elements of the cyclic subgroup generated by $(1,1)$ in the group $\mathbb{Z}_2 \times \mathbb{Z}_3$.

37.2 In-Class Exercises

Exercises

1. Determine the following values in the group $\mathbb{Z}_3 \times \mathbb{R}^*$. First, make sure you know what operations are involved in the two "factors" of this direct product.

   (a) $(2,1) * (1,2)$

   (b) the identity element

   (c) $(1, 1/2)^{-1}$

2. Describe the elements of the cyclic subgroup generated by $(1, -1)$ in the group $\mathbb{Z}^2$ as simply as possible.

3. In the text, it was observed that if $a, b \in \mathbb{R}$, then $\{(x, y) \mid ax + by = 0\}$ is subgroup of $\mathbb{R}^2$. Prove that this is true, and show that the similar set $\{(x, y) \mid ax + by = c\}$ is not a subgroup when $c \neq 0$.

4. A symmetric linked list in a five bit computer contains four nodes with the machine address of first node being 11011. The link fields for the nodes in the list from first to last are $00111, 00111, 10111, 11100$. Assume the nil pointer value is 00000. What are the machine addresses of the nodes 2, 3 and 4?

Chapter 38

Isomorphisms

38.1 Reading Assignment

Read Section 11.7 of *Applied Discrete Structures*. We've seen several different groups of order two. Two of them are the group $\mathbb{Z}_2$ and $[\{1, -1\}; \cdot]$. There is no denying that these groups are different but from an algebraic point of view, they have all the same properties. This section makes this observation more precise. Based on the definitions, these two groups (and all other groups with two elements) are *isomorphic*. In studying groups, this is essentially saying they are all equal.

Question 38.1.1 Response Question. How is the topic of isomorphisms related to the Slide Rule, p. 94, an analog computing device that was in common usage up until the 1970's. □

Also, turn in solutions to this exercise:

Exercises

1. Answer the following question: siedem razy cztery = _____?
    Hint. The question is in Polish.

2. The two groups $\mathbb{Z}_4$ and $\mathbb{U}_8$ have the same order but are not isomorphic. Give a reason why.

38.2 In-Class Exercises

Exercises

1. Write out the operation table for $G = [\{1, -1, i, -i\}; \cdot]$ where i is the complex number for which $i^2 = -1$. Show that G is isomorphic to $[\mathbb{Z}_4; +_4]$.

2. Solve $x^2 = -1$ in G by first translating the equation to $\mathbb{Z}_4$, solving the equation in $\mathbb{Z}_4$, and then translating back to G.

3. Although $\mathbb{Z}_6$ and $\mathbb{Z}_2 \times \mathbb{Z}_3$ are both groups of order six and are iso-mophic, $\mathbb{Z}_8$ and $\mathbb{Z}_2 \times \mathbb{Z}_4$ are not isomophic even though they both

have order eight. Find a reason why this is the case.

4. Think about how you might represent a file of nonzero real numbers that you plan to multiply together on computer.

    (a) Prove that $\mathbb{R}^*$ is isomorphic to $\mathbb{Z}_2 \times \mathbb{R}$.

    (b) Describe how multiplication of nonzero real numbers can be accomplished doing only additions.

38.3 The Slide Rule

The slide rule was invented in the 17th century. It was used extensively by students, scientists and engineers until the 1970's to do multiplication, division and other common arithmetic operations.

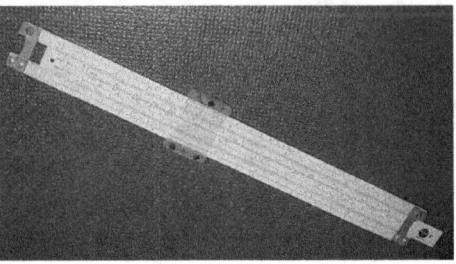

Figure 38.3.1 A Slide Rule

Chapter 39

Posets and Lattices

39.1 Reading Assignment

Read Sections 13.1 and 13.2 of *Applied Discrete Structures*. Partial orderings were introduced in Chapter 6 and you might want to review the definition before you start on Chapter 13. You might recall that partial orderings tend to compare certain pairs of elements in a poset according to some size, where size might not always mean size in the conventional sense. In any case, if we take two elements from a poset, an upper bound is something in the poset that is comparably larger than them both. A lower bound is defined similarly. In some cases, we can define binary operations ("join" and "meet") based on this idea. When we can, we get an algebraic structure called a *lattice*.

Question 39.1.1 Response Question. Google "total ordering" and find out if a total ordering is a "partial ordering." □

Also, turn in solutions to these exercises:

Exercises

1. For the poset $(\mathbb{N}, \leq)$, what are the greatest lower bound and least upper bound of two elements a and b? Are there least and/or greatest elements?

2. Let L be the set of all propositions generated by p and q. What are the meet and join operations in this lattice under implication? What are the maximum and minimum elements?

39.2 In-Class Exercises

Exercises

1. Consider the poset $(D_{50}, |)$, where $D_{50} = \{1, 2, 5, 10, 25, 50\}$.

   (a) Find all lower bounds of 10 and 25.

   (b) Find the greatest lower bound of 10 and 25.

   (c) Find all upper bounds of 10 and 25.

   (d) Determine the least upper bound of 10 and 25.

   (e) Draw the Hasse diagram for D_{50} with respect to $|$.

2.   Demonstrate that the pentagon lattice is nondistributive.

3.   We naturally order the numbers in $A_m = \{1, 2, ..., m\}$ with "less than or equal to," which is a partial ordering. We define an ordering, $\preceq$ on the elements of $A_m \times A_n$ by

$$(a, b) \preceq (a', b') \Leftrightarrow a \leq a' \text{ and } b \leq b'$$

    (a) Prove that $\preceq$ is a partial ordering on $A_m \times A_n$.

    (b) Draw the ordering diagrams for $\preceq$ on $A_2 \times A_2$, $A_2 \times A_3$, and $A_3 \times A_3$.

    (c) In general, how does one determine the least upper bound and greatest lower bound of two elements of $A_m \times A_n$, (a, b) and (a', b')?

    (d) Are there least and/or greatest elements in $A_m \times A_n$?

4.   The last question you answered with the readings for this class is actually not very precise. For example, is it true that $p \Rightarrow p \wedge (p \vee q)$ or $p \wedge (p \vee q) \Rightarrow p$, or are both true? If both are true, what does that say about the relation $\Rightarrow$? This would present a problem, but it can be fixed by considering the elements of our poset to be equivalence classes. What would be the equivalence classes?

39.3 Nondistributive lattices

Here are the fundamental nondistributive lattices.

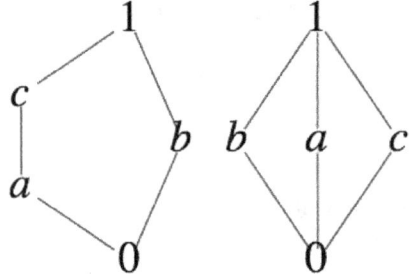

Figure 39.3.1 Nondistributive lattices, the pentagon and diamond lattices

Chapter 40

Boolean Algebras

40.1 Reading Assignment

Read Section 13.3 of *Applied Discrete Structures*. In this section we further restrict our attention to lattices that have the algebraic properties that are required of a Boolean Algebra. Propositional logic that we saw in Chapter 3 is prime model for what Boolean Algebra is like, but there several other concrete Boolean Algebras that we will consider.

Question 40.1.1 Response Question. Why can't a lattice with three elements, a mininum element, a maximum element and third element between them be a boolean algebra? □

Also, turn in solutions to these exercises:

- Let D_n be the set of positive integers that divide evenly into n. List the elements of each of the sets D_6, D_{16}, D_{12}, and D_{30}

- What is the complement of a logical proposition in the Boolean algebra of logic?

40.2 In-Class Exercises

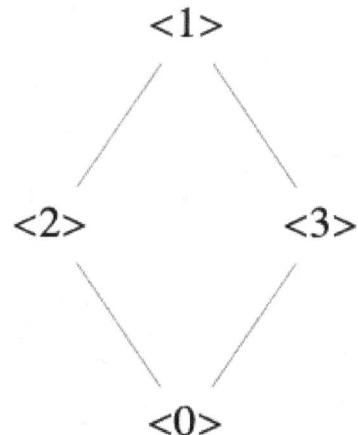

Figure 40.2.1 Lattice of subgroups of $\mathbb{Z}_6$

1. Consider the lattices of divisors of 6, 12, 16, and 30, with the partial ordering "divides".

   (a) Determine which elements of D_6 have a complement.

   (b) Determine which elements of D_{12} have a complement.

   (c) Determine which elements of D_{16} have a complement.

   (d) Determine which elements of D_{30} have a complement.

   (e) Given that D_n is always a distributive lattice, for what values of n do you suppose D_n is a boolean algebra?

2. Recall that if $[G; *]$ is a finite group and $a \in G$, the cyclic subgroup generated by a is $\langle a \rangle = \{a, a*a, a*a*a, \dots\}$, which is a finite set since you eventually get to the identity and can stop listing elements. For example, if the group is $\mathbb{Z}_6$, there are four different cyclic subgroups: $\langle 0 \rangle$, $\langle 1 \rangle$, $\langle 2 \rangle$, and $\langle 3 \rangle$. The other two elements of $\mathbb{Z}_6$, 4 and 5 generate subgroups that are equal to the subgroups generated by 2 and 1, respectively. The Hasse diagram for this set of subgroups with $\subseteq$ is Figure 40.2.1, p. 97

   (a) List the distinct cyclic subgroups generated by elements of $\mathbb{Z}_{12}$. The set of subgroups you've found is a lattice with the partial ordering $\subseteq$. Draw the Hasse diagram for this poset.

   (b) Do the same with the cyclic subgroups generated by elements of $\mathbb{Z}_{16}$.

   (c) Do the same with the cyclic subgroups generated by elements of $\mathbb{Z}_{30}$.

   (d) Any general observations from the three cases?

Chapter 41

Atoms and Tuples of Bits

41.1 Reading Assignment

Read Sections 13.4 and 13.5 of *Applied Discrete Structures*. For any positive integer, n, the number of different groups of order n can vary quite a bit. In contrast, the number of Boolean Algebras of order n is either zero or one. In these two sections, you will see why.

Question 41.1.1 Response Question. Consider the lattice of real numbers in the interval $[0, 1]$ with the relation $\leq$. Does this lattice have any atoms? □

Also, turn in solutions to these exercises:

- What are the atoms of the lattice of subsets of $\{1, 2, 3, 4, 5, 6, 7\}$ generated by the two sets $A_1 = \{1, 3, 5, 7\}$ and $A_2 = \{1, 2, 3, 5\}$ with the partial ordering $\subseteq$?

- What are the atoms of the lattice of positive integers with the relation "divides?"

41.2 In-Class Exercises

1. (a) There are many different, yet isomorphic, Boolean algebras with two elements. Describe one such Boolean algebra that is derived from a power set, $\mathcal{P}(A)$, under $\subseteq$. Describe a second that is described from D_n, for some $n \in P$, under "divides."

   (b) Since the elements of a two-element Boolean algebra must be the greatest and least elements, 1 and 0, the tables for the operations on $\{0, 1\}$ are determined by the Boolean algebra laws. Write out the operation tables for $[\{0, 1\}; \vee, \wedge, -]$.

2. Consider the Boolean algebra $\mathcal{B} = [B_2^6; \vee, \wedge,]$, and let E_6 be the set of bit strings of length six having an even number of 1's. Is E_6 closed with respect to the operations of $\mathcal{B}$.

3. Give an example of a Boolean algebra of order 16 whose elements are certain subsets of the set $\{1, 2, 3, 4, 5, 6, 7\}$, the minimum element is the empty set, and the maximum element is $\{1, 2, 3, 4, 5, 6, 7\}$.

4. Give an example of a Boolean algebra of order 16 whose elements are a certain set of bit strings of the set B_2^7, whose minimum element is 0000000, and whose maximum element is 1111111.

Chapter 42

Cyclic Groups

42.1 Reading Assignment

Read Sections 15.1 of *Applied Discrete Structures*. Some groups have the property that they are formed as a cyclic subgroup of one of its elements. When this is the case, the group itself has a particularly simple form, which we examine in this section.

Question 42.1.1 Response Question. Google "Chinese Remainder Theorem". Why does the theorem have this name? □

Also, turn in solutions to these exercises:

- List all of the distinct cyclic subgroups of the cyclic group $\mathbb{Z}_{15}$. Explain why this list is actually a list of *all* of the subgroups of $\mathbb{Z}_{15}$.

- What are the generators of the multiplicative group $\{5^k \mid k \in \mathbb{Z}\}$?

42.2 In-Class Exercises

1. What are the generators of the group $\mathbb{Z}_{15}$? What relationship do these numbers have with the modulus, 15?

2. Suppose $[G; *]$ is a cyclic group with generator g. If you build a graph of with vertices from the elements of G and edge set $E = \{(a, g*a) \mid a \in G\}$, what would the graph look like? If G is a group of even order, what would a graph with edge set $E' = \{(a, g^2 * a) \mid a \in G\}$ look like?

3. The function $f : \mathbb{Z}_{77} \to \mathbb{Z}_7 \times \mathbb{Z}_{11}$ defined by $f(a) = (a \bmod 7, a \bmod 11)$ is an isomorphism. Suppose you are given some elements in $\mathbb{Z}_{77}$ and you want there sum (mod 77). Furthermore, suppose you mapped these numbers, using f, to $\mathbb{Z}_7 \times \mathbb{Z}_{11}$ and there sum in the direct product was $(1, 6)$. What is the sum in $\mathbb{Z}_{77}$?

Chapter 43

Cosets

43.1 Reading Assignment

Read Section 15.2 of *Applied Discrete Structures*. If you get confused in this section, think about the partition of the integers into even integers and odd integers. There is a consistency in doing arithmetic on the integers where, for example, any time you add an even integer and an odd integer, you get an odd integer. So even + odd = even. This is coset addition! For any group, the only way something like this can work is when the partition is into cosets, which is why we start by defining this set of subsets of a group.

Question 43.1.1 Response Question. How did Galois die, and why am I asking you this? □

Also, turn in solutions to these exercises:

1. What does Lagrange's Theorem say about the possible subgroups of a group of order 12?

2. Let H be the subgroup $\{0, 25, 50, 75\}$ of the additive group $\mathbb{Z}_{100}$. How many distinct left cosets are there of H? What coset contains the number 42? List its elements.

43.2 In-Class Exercises

1. List the distinct left cosets of the cyclic subgroup of U_{13} generated by 3. To what other group is the factor group of these cosets isomorphic?

2. For each group and subgroup, to what simpler group is G/H isomorphic? Draw a picture of the groups and their cosets.

   (a) $G = \mathbb{R}^2$ and $H = \{(a, a) \mid a \in \mathbb{R}\}$

   (b) $G = \mathbb{Z}_2^3$ and $H = \langle 111 \rangle$. In this group, write triples as bit strings of length 3.

3. Consider the subgroup $C = \{00000, 01101, 11110, 10011\}$ of the group $S = \mathbb{Z}_2^5$. Suppose your answer to a multiple choice question with four possible responses is stored in five bits of computer memory using one of the four elements of C. Now suppose the computer is hacked but we know that no more than one of the five bits in your response has been changed. Can your response be recovered?

Chapter 44

Permutation Groups, Part 1

44.1 Reading Assignment

Read the first two subsections of Section 15.3 of *Applied Discrete Structures*, stopping when you get to Subsection 15.3.3. Our focus in this section is on whole sets of functions, but we also consider convenient ways to represent individual functions. In most cases, you'll find that cycle notation is probably the best option, even though it might look strange at first.

Question 44.1.1 Response Question. Joseph-Louis Lagrange (1736-1813) first thought of permutations as functions from a set to itself. However, he didn't invent cycle notation. Who did that?   □

Also, turn in solutions to these exercises:

1 Write the following elements of S_5 as a product of disjoint cycles.

(a) $\alpha = \begin{pmatrix} 1 & 2 & 3 & 4 & 5 \\ 2 & 4 & 1 & 5 & 3 \end{pmatrix}$ (b) $\beta = \begin{pmatrix} 1 & 2 & 3 & 4 & 5 \\ 4 & 2 & 5 & 1 & 3 \end{pmatrix}$

2 Compute the following using the values of α and β above. Write your answer as a product of disjoint cycles.

(a) $\alpha \circ \beta$ (c) α^{-1}

(b) $\beta \circ \alpha$ (d) β^{-1}

44.2 In-Class Exercises

There are advantages to cycle notation. One is that the order of an element can be determined by its cycle structure.

1. What are the orders of the following permutations?

(a) $(1,2,3)$ (c) $(1,2,3)(4,5)$ (e) $(1,2,3,4)(5,6)$

(b) $(1,2,3,4)$ (d) $(1,2,3)(4,5,6)$ (f) $(1,6)(1,5)(1,4)(1,3)(1,2)$

2. Based on the result in the previous problem, predict the orders of the following permutations without actually computing their powers.

    (a) $(1, 2, 3, 4, 5, 6, 7)$ (b) $(1, 2, 3, 4)(5, 6, 7, 8)$ (c) $(1, 2, 3, 4)(5, 6, 7)$

3. Among all permutations in S_{10}, what is their largest order? How about S_{11}?

4. Prove that if σ and ϕ are permutations, then the order of $\phi \circ \sigma \circ \phi^{-1}$ is equal to the order of σ.

44.3 The order of a group element

Here is a reminder of the definition of *order*.

Definition 44.3.1 Order of a Group Element. The order of an element a of group G is the number of elements in the cyclic subgroup of G generated by a. The order of a is denoted $ord(a)$. $\Diamond$

Chapter 45

Permutation Groups, Part 2

45.1 Reading Assignment

Read the rest of Section 15.3 of *Applied Discrete Structures*. The family S_n of symmetric groups is extremely important in group theory, particularly when it comes to doing computations. There is a theorem called Cayley's Theorem that essentially says that *every* group is subgroup of some symmetric group. This means that if you intend to do group computations, it's natural to implement the work in terms of permutations.

Question 45.1.1 Response Question. How is Rubik's cube related to permutation groups? □

Also, turn in solutions to these exercises:

- Based on *Lagrange's Theorem*, what are the possible orders of subgroups of the group S_4?

- The *shape* of a permutation, σ in S_n is the sequence of lengths of the cycles in the representaion of σ as a product of disjoint cycles. The lengths are written in descending order. For example, if $\sigma = (1, 3, 4)(8, 11)(2, 5, 6, 7)(9, 10)(12) \in S_{12}$, then its shape is 4,3,2,2,1. List all possible shapes of elements in S_4.

45.2 In-Class Exercises

1. Prove that $H = \{\sigma \in S_5 \mid \sigma(5) = 5\}$ is a subgroup of S_5. How many elements does H have?

2. Given a graph with vertex set $\{1, 2, 3, 4, 5\}$, the group of automorphisms of that graph will be a subgroup of the group S_4. For each of the following subgroups of S_5, draw a graph (directed or undirected) that has that subgroup as its automorphism group.

   (a) $\{i\}$

   (b) The cyclic subgroup generated by $(1, 2)$

   (c) The cyclic subgroup generated by $(1, 2, 3, 4, 5)$

   (d) $\{\sigma \in S_5 \mid \sigma(5) = 5\}$

3. There are subgroups of S_n that are not the automorphism group of of any graph with n vertices. Illustrate this fact by showing that it is true for

S_4. However, for any such subgroup, H, there are graphs with a larger number of vertices that have automorphism groups that are isomorphic to H. Illustrate this fact with the subgroup you identified in S_4.

Chapter 46

Algebraic Coding, Part 1

46.1 Reading Assignment

Read the definition of *Homomorphism* in Section 15.4. Then read Section 15.5 of *Applied Discrete Structures* up to, but not including, Subsection 15.5.2 Error Correction. After outlining the objectives behind coding theory we first consider how to effectively detect transmission errors.

Question 46.1.1 Response Question. If you measure the distance between elements of a set, S, with a function $d : S \times S \to \mathbb{R}$, what general distance properties would you expect of this function? ☐

Also, turn in solutions to these exercises:

1. *Interogating a Liar.* I'm a liar, but not a big one. In my responses to your yes/no questions, I promise not to lie more than one time. I'm thinking of a number, either 0 or 1. How many yes/no questions do you need to ask in order to find out my number? Explain your logic.

2. The *Hamming Distance*, d_H, between two strings of bits with the same length is the number positions within the two strings where the strings differ. For instance, $d_H(11101, 10100) = 2$ because the two strings differ in positions 2 and 5. What is the minimal Hamming distance between any two strings in the set

$$C = \{00000, 11011, 11100, 00111\}?$$

46.2 In-Class Exercises

1. Suppose a two bit message is encoded into a five bit message using the function $e(b_1 b_2) = b_1 b_1 (b_1 +_2 b_2) b_2 b_2$. What matrix, G, has the property that $e(b_1 b_2) = (b_1 b_2) G$?

2. Define the two functions $g : \mathbb{Z}_2{}^3 \to \mathbb{Z}_2{}^4$ and $p : \mathbb{Z}_2{}^4 \to \mathbb{Z}_2$ by $g(a_1, a_2, a_3) = (a_1, a_2, a_3, a_1 +_2 a_2 +_2 a_3)$, and $p(b_1, b_2, b_3, b_4) = b_1 + b_2 + b_3 + b_4$.

   (a) The values of $g(a_1, a_2, a_3)$ and $p(b_1, b_2, b_3, b_4)$ can be computed as matrix products. Determine the matrices.

   (b) Describe the function $p \circ g$. Is it a homomorphism?

3. Suppose we were to send three bits across a binary symmetric channel by first encoding the bits using the function g from the previous problem, producing a sequence of four bits. After receiving the four bits, the person at the receiving end applies the function p to those bits. What can we derivive form the output of p?

4. *Linear Codes.* Let m and n be positive integers. Suppose that G is an $m \times n$ matrix of zeros and ones. Then we can can define a function $e : \mathbb{Z}_2^m \to \mathbb{Z}_2^n$ by $e(a) = aG$ where we take $a \in \mathbb{Z}_2^m$ to be a $1 \times m$ matrix and we perform mod 2 arithmetic in computing the product aG. Prove that e is a homomorphism.

Chapter 47

Algebraic Coding, Part 2

47.1 Reading Assignment

Read the rest of Section 15.5 of *Applied Discrete Structures*. We consider how to correct transmission errors by using redundant information in a coded message. This avoids having to ask for a message to be resent from the sender.

Question 47.1.1 Response Question. Who was Claude Shannon and what is his connection with coding theory? □

Also, turn in solutions to these exercises:

1. What is the minimum Hamming distance between code words in the set $\{11100, 00111, 10001, 01010\}$? Without doing any calculation, how can you tell

$$\{11100, 00111, 10001, 01010\}$$

   cannot be the code words of a linear code?

2. I'm a liar, but not a big one. In my responses to your yes/no questions, I promise not to lie more than one time. My favorite number is 0, 1, 2, or 3. How many yes/no questions do you need to ask in order to figure it out? You need to submit all of your questions to me without waiting for my answers. (thanks to Daniel Glasscock for this problem)

47.2 In-Class Exercises

1. Suppose a two bit message is encoded into a five bit message using the function $e(b_1 b_2) = b_1 b_1 (b_1 +_2 b_2) b_2 b_2$. Determine a matrix P such that if a two bit message, $b = b_1 b_2$, is encoded and $e(b)$ transmitted, then any single bit error in the received string $c = c_1 c_2 c_3 c_4 c_5$ can be identified by whether cP is the zero vector or not.

2. Consider the linear code defined by the generator matrix

$$G = \begin{pmatrix} 1 & 0 & 1 & 0 \\ 0 & 1 & 1 & 1 \end{pmatrix}$$

   (a) What size blocks does this code encode and what is the length of the code words?

   (b) What are the code words for this code?

(c) With this code, can you detect single bit errors? Can you correct all, some, or no single bit errors?

3. To build a **rectangular code**, you partition your message into blocks of length m and then factor m into $k_1 \cdot k_2$ and arrange the bits in a $k_1 \times k_2$ rectangular array as in the figure below. Then you add parity bits along the right side and bottom of the rows and columns. The code word is then read row by row.

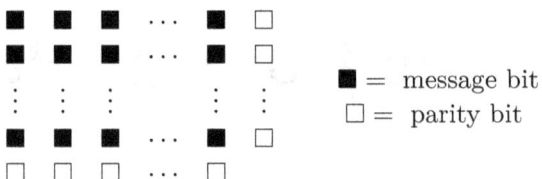

$\blacksquare$ = message bit
$\square$ = parity bit

For example, if m is 4, then our only choice is a 2 by 2 array. The message 1101 would be encoded as

$$
\begin{array}{cc|c}
1 & 1 & 0 \\
0 & 1 & 1 \\
\hline
1 & 0 &
\end{array}
$$

and the code word is the string 11001110.

(a) Suppose that you were sent four bit messages using this code and you received the following strings. What were the messages, assuming no more than one error in the transmission of coded data?

   (i) 11011000   (ii) 01110010   (iii) 10001111

(b) If you encoded n^2 bits in this manner, what would be the rate of the code?

(c) Rectangular codes are linear codes. For the 3 by 2 rectangular code, what are the generator and parity check matrices?

4. A code with minimum distance d is called *perfect* if every string of bits is within Hamming distance $r = \frac{d-1}{2}$ of some codeword. For such a code, the spheres of radius r around the codewords partition the set of all strings. This is analogous to packing objects into a box with no wasted space. Using just the number of bit strings of length n and the number of strings in a sphere of radius 1, for what values of n is it possible to find a perfect code of distance 3? You don't have to actually find the codes.

References

[1] Bogart, Kenneth P., *Combinatorics Through Guided Discovery*, http://bogart.openmathbooks.org

[2] Doerr, A, and K. Levasseur, *Applied Discrete Structures*, http://discretemath.org

[3] Levin, Oscar, *Discrete Mathematics: An Open Introduction*, http://discrete.openmathbooks.org,
A one semester open source text with some nice features

[4] David Pengelley, *From Lecture to Active Learning: Rewards for All, and Is It Really So Difficult?*, The College Math Journal, January 2020, **51** no. 1, 13–24, doi.org/10.1080/07468342.2020.1680228
The idea of teaching in an active learning environment had been on my radar for a while and I'd experimented with some aspects of the format. This article was the final impetus for launching this project.

Index

www.ingramcontent.com/pod-product-compliance
Lightning Source LLC
Chambersburg PA
CBHW080936180526
45158CB00024B/2193